Cloud Computing for Science a

Scientific and Engineering Computation

William Gropp and Ewing Lusk, editors; Janusz Kowalik, founding editor

A complete list of books published in the Scientific and Engineering Computation series appears at the back of this book.

/

Cloud Computing for Science and Engineering

Ian Foster and Dennis B. Gannon

The MIT Press

Cambridge, Massachusetts

London, England

This book was set in LATEX by the authors. Printed and bound in the United States of America.

Library of Congress Cataloging-in-Publication Data is available.

ISBN: 978-0-262-03724-2

10 9 8 7 6 5 4 3 2 1

Contents

Website

Cloud4SciEng.org

This book is accompanied by the above website. This site provides a variety of supplementary material, including exercises, lecture slides, corrections of mistakes, and other resources that should be useful to both readers and instructors.

Acknowledgments

This book has benefited from the contributions of many people. We first thank Rusty Lusk for persuading us to write the book and Marie Lufkin Lee, our editor at MIT Press, for her support as we developed the manuscript.

A special thanks to Rich Wolski and Stig Telfer who, at our invitation, wrote the chapters on Building your own Cloud with Eucalyptus and OpenStack, respectively. We would not have part IV of the book without their contributions.

We thank Ben Blaiszik, Kyle Chard, Ryan Chard, Ricardo Barros Lourenço, Jim Pruyne, Tyler "Eagle Eye" Skluzacek, Roselyne Tchoua, and Logan Ward for their careful review and detailed critiques of many chapters, and Gail Pieper for her masterly copyediting help, which as in many past projects, turned our often turgid and sometimes erroneous text into prose.

We also thank Pete Beckman and Charlie Catlett for providing information on the Array of Things and Manish Parashar for information about the Ocean Observatory Initiative, both featured in chapter 9; Rachana Ananthakrishnan, Kyle Chard, Eli Dart, Steve Tuecke, and Vas Vasiliadis for their contributions to the research data portal described in chapter 11; Ravi Madduri for his contributions to the material on Globus Genomics in chapter 14; Stephen Rosen for providing much of the material on the Globus service in chapter 14; and Ben Blaiszik for testing all of the Jupyter notebooks featured in chapter 17. Tyler Skluzacek provided a unique perspective on Resnet-152.

We are grateful to Ian Goodfellow, Yoshua Bengio, and Aaron Courville, authors of the magnificent *Deep Learning* [143], for providing their LaTeX macros and introducing us to pdf2htmlEX, which we used to make the web version of the book. Thanks also to Lu Wang for writing pdf2htmlEX.

We owe a deep vote of thanks to the U.S. Department of Energy, National Science Foundation, National Institutes of Health, and National Institute of Standards and Technology for the support they have provided to our research over many years: support that has allowed us to gain the knowledge of cloud and technical computing that we have worked to distill in this book. Not just this book but modern science, and indeed the cloud that we describe in the chapters that follow, would not exist without the vision and persistence that have allowed these agencies to support research excellence for many decades. We are also grateful to Amazon and Microsoft for generous grants of cloud time.

Argonne National Laboratory and the University of Chicago provided the first author with a uniquely collegial and stimulating environment in which to undertake this work, as they have for many years.

Finally, we acknowledge that the book is only possible because of the work of the many talented and dedicated architects and developers who produced the tremendous array of software that makes cloud computing so rewarding and fun.

Preface

"Civilization advances by extending the number of important operations
which we can perform without thinking about them."
—Alfred North Whitehead, *Introduction to Mathematics*

This book is about cloud computing and how you, the reader, may apply it to advantage in science and engineering. It is a practical guide, with many hands-on examples, all accessible online, of how to use cloud computing to address specific problems that arise in technical computing, and actionable advice on how and when to apply cloud computing in your daily work.

The term cloud computing was first used in 1996; today, it appears on billboards in airports. You may wonder whether it is a technology, movement, repackaging of old ideas, or marketing slogan. It is all of these things and more. Above all, though, it is a transformation in the state of affairs that we ignore at our peril, in science as in other parts of our life. As Tim Bray wrote in 2015: "Yeah, computing is moving to a utility model. Yeah, you can do all sorts of things in a public cloud that are too hard or too expensive in your own computer room. Yeah, the public-cloud operators are going to provide way better uptime, security, and distribution than you can build yourself. And yeah, there was a Tuesday in last week." [77]

This emergence of powerful, always-on, accessible cloud utilities has transformed how we, as consumers, interact with information technology, allowing us to stream videos from Netflix (hosted on the Amazon cloud), search for web content via Google (leveraging the Google cloud), update friends on our doings via Facebook, and ask Alexa to buy our groceries. The cloud has also allowed many companies to outsource much of their information technology to cloud providers, slashing costs and increasing velocity. Myriad previously manual activities are being automated via software running on cloud utilities, in ways imagined by McCarthy in 1960 and explored by grid in the 1990s, and now realized at scale by cloud providers such as Amazon, Google, and Microsoft.

But what about science and engineering? Many scientists and engineers use cloud services such as Dropbox, GitHub, Google Docs, Skype, and even Twitter in their work. But they are far from exploiting the full benefits of cloud computing. Some technical applications run on cloud computers, but few researchers outsource much else to the cloud. This is a missed opportunity. After all, science and engineering, while fascinating and intellectually rewarding professions, include many mundane and time-consuming activities. Can we not accelerate discovery (and have more fun) via automation and outsourcing? We believe that the answer to this question is yes, which is why we wrote this book.

In the chapters that follow, we examine the new technologies that underpin cloud, the new approaches to technical problems that cloud enables, and the new ways of thinking that are required to apply cloud effectively in research. We do not aspire to provide a comprehensive guide to cloud computing: the major cloud providers operate literally hundreds of services, and there are surely many beyond those presented here that can be applied effectively in science and engineering. But we do describe the essentials and provide you with the concepts required to integrate cloud services into your work.

The following are some of the questions that we find people asking, and for which we aim to provide answers. Should I buy a cluster or use cloud? Will my grant pay the bills if I use a commercial cloud? How can I get my data to the cloud? Is it safe there? Can I share it with my collaborators? How do I compute in the cloud? Can cloud computing scale? What if I want to compute on large quantities of data? Should I use cloud platform services in my work? Which ones are good for science and engineering? How can I build my own cloud services? Can I make them scale on demand to address really large problems? What are some examples of successful uses of cloud in science and engineering? How can I build my own cloud?

Lacking a crystal ball, we cannot provide definitive answers to these questions. But we can at least provide you with information and perspectives that you can use to make up your own mind on these and other questions.

 All flows, nothing stays. So wrote Heraclitus 2,500 years ago, and software is worse. Some technical details in this book will prove more transient than we would like. Do not despair. Help us and your colleagues by letting us know at `Cloud4SciEng.org`. We will update the website, and prepare for the second edition.

Chapter 1

Orienting in the Cloud Universe

"I've also grown weary of reading about clouds in a book. Doesn't this piss you off? You're reading a nice story, and suddenly the writer has to stop and describe the clouds. Who cares?"

—George Carlin, "Seven Things I'm Tired Of"

We start this journey into cloud computing for science and engineering by introducing important concepts and the structure of the book, and reviewing tools that you should know in order to obtain the most value from this material.

1.1 Cloud: Computer, Assistant, and Platform

Scientists and engineers can apply cloud capabilities in their work in many different ways. We find it useful to think in terms of three categories of use.

First, a cloud is an **elastic computer**: a source of on-demand computing and storage that you can call upon when you need computing or storage capacity larger than, or different from, what is available locally. Accessing this capacity in the cloud may be cheaper, faster, and/or more convenient than acquiring and operating your own computing and storage systems. While there are differences between the cloud computing and storage offerings from different cloud providers, they provide quite similar capabilities: in particular, object storage and execution of virtual machines and containers. We cover this **infrastructure as a service** (IaaS) technology and its applications in parts I and II.

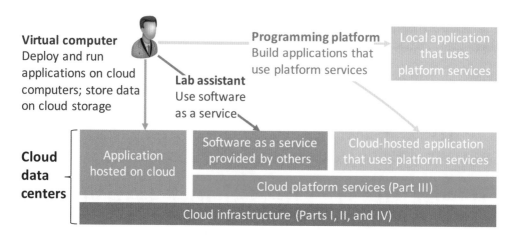

Figure 1.1: Scientists can use clouds in three distinct ways: As a source of on-demand computing and storage on which to run their own software (left); as a source of software that can be run over the network (center) as a source of new platform capabilities that can allow development of new types of software (right).

Second, a cloud is a tireless **laboratory assistant**: a source of powerful software that can perform certain tasks more effectively and/or cheaply than you can yourself: for example, Academia.edu, Google Scholar, and ResearchGate to access information about publications, facilitating research and citation; GitHub to manage software and documents, facilitating collaboration, software sharing, and reproducibility; Google Docs, Box, and Dropbox to share data; Science Exchange to order experiments; Figshare for publishing data; Globus to move and manage large data; Skype and other services for communication; and many others. In each case, you can avoid substantial cognitive, administrative, and financial burdens that you or members of your laboratory would incur if they had to perform these tasks themselves. These **software as a service** (SaaS) capabilities are important, but are largely out of scope for this book, although we do discuss how to build your own software as a service in chapter 14.

Third, a cloud is a **programming platform**: that is, a collection of powerful software mechanisms that you can use to build software with capabilities that would be difficult or expensive to duplicate in your own lab: for example, an event processing system that can process millions of events per second, a database that can scale to billions of rows, an identity management service that can handle dozens of different identity providers, a data transfer service that can move terabytes securely and reliably, or a service that is replicated in multiple geographic regions to ensure continuous operations. These platform capabilities are arguably the most exciting part of cloud computing, because they enable individual programmers to create and operate software systems that would otherwise require large teams. They allow the cloud to be used as an interactive environment for large-scale computational experimentation and discovery. They can also be the most challenging to use effectively, because they have often been developed for use cases rather different from traditional technical computing. In addition, it is in this area that we see the biggest variation across cloud vendors in terms of capabilities and interfaces. We discuss these **platform as a service** (PaaS) capabilities in part III.

Inevitably, the boundaries between these different types of cloud system and usage are not always crisply defined. For example, a growing number of software-as-a-service offerings provide APIs that allow them to be used as platform services, and we often see (as discussed in part III) platform services enhancing the value of virtual computer offerings.

1.2 The Cloud Landscape

The cloud landscape is large, diverse, and complex. The U.S. National Institute of Standards and Technology (NIST) lists five essential characteristics of cloud computing: on-demand self-service, broad network access, resource pooling, rapid elasticity or expansion, and measured service [197]. Today, thousands of companies offer services with some or all of these characteristics, from low-level computing and storage to sophisticated software: see figure 1.2. But apart from the collaboration and content management systems listed above, few of the commercial cloud services shown in the figure are relevant to science and engineering.

One major exception is in the realm of cloud infrastructure: the elastic compute services that allow individuals to acquire storage and computing on demand. Here, the landscape is simpler, particularly when we focus on providers with offerings relevant to science and engineering. (Others specialize in specific products, such as Oracle for databases or AT&T for telecom.) Three vendors, Amazon, Google, and Microsoft, dominate the industry, as shown in table 1.1 on page 5, and each has proven useful for science and engineering. We focus in this book on the services provided by those three providers and by one academic research cloud, Jetstream [122] `jetstream-cloud.org`*. Nevertheless, other cloud providers are also impressive. For example, the New York-based DigitalOcean is popular in the software engineering and cloud application development community, while Rackspace supports those using the Amazon and Microsoft clouds and also runs its

---

*A shaded, rounded rectangle denotes an `https` URL, in this case `https://jetstream.org`.

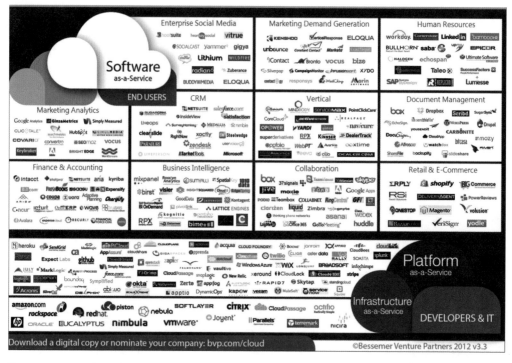

Figure 1.2: While dated, Bessemer Venture Partner's picture of the top 300 cloud computing companies in 2012 conveys the vast range of cloud service providers.

own cloud servers. European cloud providers include 1&1, UpCloud, City Cloud, CloudSigma, CloudWatt, and Aruba Cloud. Large telecommunications and search companies, such as China's Baidu, are also rapidly building cloud data centers. Together, these various companies operate more than one hundred data centers around the globe, containing an estimated ten million servers and vast storage. (We base these estimates on news articles [63, 169, 96, 204].)

The cloud services operated by Amazon, Google, and Microsoft are commonly referred to as **public clouds**, by analogy with the public utilities (power, telephone, water, sewer, etc.) on which most of us depend in our daily lives. Like public utilities, they provide computing, storage, and other services to any member of the public with the ability to pay. (Public clouds are not regulated like public utilities, leading some to argue that the term is inappropriate.)

In contrast, a **private cloud** is operated by a private institution or individual to provide computing, storage, and/or other services to a more limited audience. We can think of them as being analogous to a private electricity generator, although the utility that they provide is not electricity but on-demand access to computing,

4

Table 1.1: Major cloud infrastructure providers of relevance to research.

Amazon	Market leader. Computing, storage, and platform services. Extensively used in science and engineering.
Microsoft	Second biggest player. Computing, storage, and platform services with both individual and enterprise customers.
Google	Began with a service called App Engine and is now using that experience to release a full suite of cloud capabilities.

storage, or software. Private clouds are frequently deployed in larger enterprises. IBM, VMware, and Microsoft are major providers of proprietary solutions for building on-premises cloud-like systems. OpenStack `openstack.org` is the dominant open source cloud software solution, particularly in the US; this software is used, for example, by Jetstream and in some public clouds, such as that operated by Rackspace. OpenNebula [202] `opennebula.org` is another prominent open source solution, used extensively in Europe. The European Union has announced plans for a Europe-wide science cloud [163] by 2018 `hnscicloud.eu`. There are also numerous academic cloud projects in Europe.

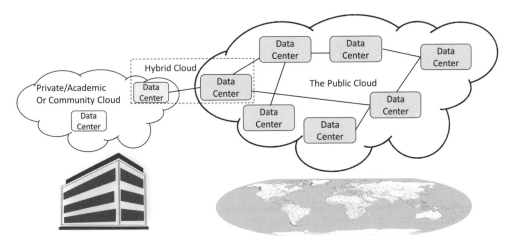

Figure 1.3: Private (including academic and community), public, and hybrid clouds.

This distinction between public and private clouds may appear minor, but it has important implications. Because the major public clouds operate at a scale far larger than any private cloud, they can offer a broad set of powerful features: for example, elasticity, fine-grained billing, high reliability due to geographic distribution, a wide variety of resource types, and rich sets of platform services. Equally importantly, they can achieve substantial economies of scale.

5

Private clouds, in contrast, typically offer a limited set of cloud-like capabilities: for example, the ability to deploy virtual machine instances and store objects. As we will see in parts I and II of this book, these capabilities are enough to support interesting applications. However, the lack of the many other services offered by the Amazon, Microsoft and Google platforms limits the range of things that can be done. Private clouds may nevertheless be preferred in some circumstances, for example because a specific workload can be run more cost-effectively on an in-house infrastructure, or because a company or researcher does not wish sensitive data to leave their premises. In such cases, so-called **hybrid clouds** may be used to run selected tasks on public clouds: a process termed **cloud bursting**. (Cloud computing has inspired some terrible terminology!)

A **community cloud** is a private cloud deployment designed to support a specific community: for example, the genomics community or a set of companies or academic institutions who want to share resources. The term **academic cloud** is sometimes used to refer to a private or community cloud focused on the needs of the academic community. Figure 1.3 depicts these different cloud types.

Private or public? The merits of private clouds are hotly debated. Proponents of private clouds argue that it can be significantly cheaper to acquire and operate dedicated computers and storage systems than to buy time on a public cloud. To give just one example, let's consider the problem of providing online access to a petabyte of data. Storing that petabyte for a year in the Amazon public cloud object store would cost, as of January 2017, $252,000 in the Amazon US East region. In contrast, you can buy a 1 PB capacity SpectraLogic Verde system, a high-capacity storage device, for $75,000, and that system of course should be usable for several years.

A second frequently cited reason for using a private cloud is the need to protect sensitive data. Increasingly, such data *can* be stored on public cloud resources from a regulatory and policy perspective, as we discuss in chapter 15, but again the costs can be daunting, particularly if your institution provides you with secure storage and computing at subsidized rates. (If they do not, then the public cloud can enable research that you could not undertake otherwise.)

Critics respond that private cloud enthusiasts underestimate the costs associated with creating and running a cloud computing system, the difficulties inherent in achieving high reliability and security, and the benefits of a truly elastic cloud that always has available capacity. (Returning to your petabyte, who pays for power, space, operations, and support? What about backups? And if you don't need online access, Amazon provides an archival storage service, **Glacier**, that can store that same petabyte for just $48,000 for a year, with automated migration of infrequently used data from object store to archive.) The question of whether to build or buy is complex, with the answer depending on many factors. Suffice to say that you should be careful to consider all relevant factors when choosing your cloud solution.

We focus primarily in this book on public clouds, as they tend to be more capable, more accessible, and easier to use than other clouds. However, we do include material on the Eucalyptus and OpenStack software that are commonly used to create private, community, and academic clouds, and on the Jetstream academic cloud. Table 1.2 lists some private clouds from the academic community.

Table 1.2: Some private research clouds and their characteristics.

Name	Description
Aristotle	Hybrid cloud for academic research, integrating Eucalyptus private cloud clusters and public cloud providers. `federatedcloud.org`
Bionimbus	A cloud-based infrastructure for managing, analyzing and sharing genomics datasets. `bionimbus.opensciencedatacloud.org`
Chameleon	A configurable experimental environment for large-scale cloud research. `chameleoncloud.org`
Jetstream	Cloud computing for the U.S. academic community, operated as part of the XSEDE research network. `jetstream-cloud.org`
RedCloud	Subscription-based cloud that provides virtual servers and storage on demand. `cac.cornell.edu/redcloud/`

1.3   A Guide to This Book

This book has been written with you, the student, in mind. (Even if you are a senior scientist or engineer, we know that you are still a student at heart!) Your discipline may be physics, astronomy, biology, engineering, computer science, the humanities, or one of the newer disciplines called computational or data science. You may have come to this book because you have heard of new ways of computing in the cloud and want to learn whether they matter to you. Perhaps:

- you have a lot of data that must be analyzed by remote collaborators;

- your current computing platform (e.g., your laptop) is no longer big enough for your needs, and you lack access to a large cluster or supercomputer;

- you have access to a supercomputer, but it does not work well for interactive data analysis and collaboration tasks;

- you want to apply new computational methods, such as machine learning or stream analytics, that are hard to install, operate, and scale; or

- you want to make software or data available to your community as a service.

We organize this book into five parts (see figure 1.4 on the following page), covering the following topics:

1. **Managing data in the cloud**: We describe the various types of data storage systems that are available for use in the cloud, and illustrate how you can interact with these services using a cloud portal or directly with code.

2. **Computing in the cloud**: Here we explore the spectrum of cloud computing capabilities. These range from deploying single virtual machines or containers for simple interactive computing to clusters of machines for data analytics or traditional high performance computing.

3. **Cloud as platform**: Beyond data storage and computing there are high-level services that are particularly well suited to research applications. We examine data analysis, machine learning, and streaming data analysis methods. We also look at some specialized cloud tools designed specifically for science.

4. **Building your own cloud**: It is possible to build a basic cloud from scratch using some powerful open source software packages. We describe two examples and some of the tools needed.

5. **Security and other topics**: Security is always a major concern for any online activity. We address this topic at the end of the book, not because it is unimportant but because managing security requires an understanding of cloud architecture, as presented in previous chapters. We also consider some concerns and thoughts about future cloud evolution.

1.4 Accessing the Cloud: Web, APIs, and SDKs

We have explained how the cloud can be used variously as a virtual computer, assistant, or platform. But how exactly do you use it for each of these things? We provide details in later chapters, but let us first explain some basic concepts.

1.4.1 Web Interfaces, APIs, SDKs, and CLIs

Most cloud services can be accessed in multiple ways. First, most support access via the web, thus permitting intuitive point and click access without any programming or even local software installation (beyond a web browser) on your part. The availability of such intuitive interfaces is part of the attraction of cloud services.

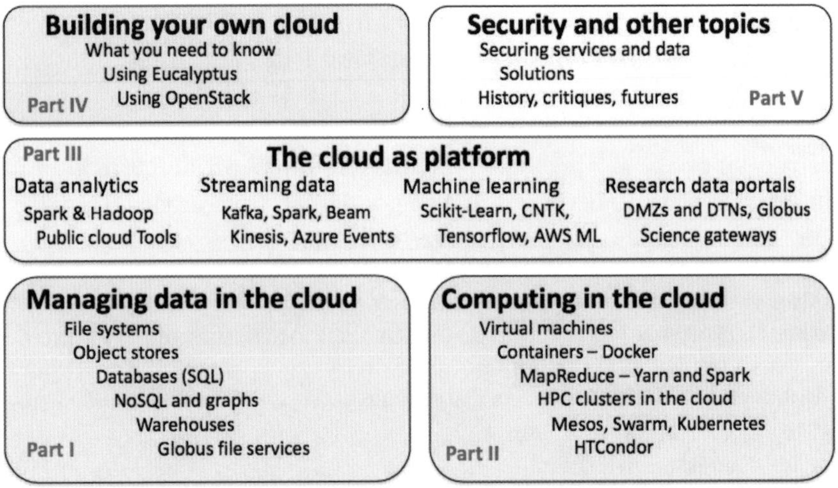

Figure 1.4: The cloud for science, from the ground up.

A web interface becomes tedious if the same or similar actions must be performed repeatedly. In such cases, you likely want to write programs that issue requests to cloud services on your behalf. Fortunately, most cloud services support such programmatic access. Typically, they support a **Representational State Transfer** (REST) application programming interface (API) that permits requests to be transmitted via the secure Hypertext Transfer Protocol (HTTPS) that is used by web browsers. (This common use of HTTPS is not a coincidence: the web interfaces discussed in the first paragraph are often implemented via browser-hosted Javascript programs that generate such REST messages.) REST APIs are the key to programmatic interactions with cloud services.

> **The meaning of REST**. This term was introduced by Roy Fielding in 2000 [121], who defined a set of principles that should be followed to build distributed systems that have desirable properties of the World Wide Web, such as performance, reliability, scalability, and simplicity. These principles define that, among other things, a REST (or RESTful) web service should refer to objects by uniform resource identifiers, such as `myserver.org/myobject`, and that operations on these objects should be performed via HTTP operations, with for example a PUT being usually interpreted as a request to create an object and a GET as a request to access its contents.

One way to interact with cloud services programmatically is to write programs that generate REST messages directly. However, while constructing REST messages "by hand" may appeal to hard-core system programmers, you will normally want to access cloud services via **software development kits** (SDKs) that you install on your computer. Such SDKs permit access from programming languages such as Python (our choice in this book), C++, Go, Java, PHP, and Ruby. (Sorry, Fortran programmers, but Fortran SDKs are few and far between.) They typically render operations on cloud services in ways that are consistent with the programming model of the language in question. Cloud vendors typically provide SDKs for accessing their services, but there are also good open source ones available, and if you do not like any of them, you are free to develop your own.

Accessing a cloud service. We use a simple example to illustrate these different approaches to accessing cloud services. Consider the Amazon Simple Storage Service (S3), which as we describe in chapter 2, allows you to create and access containers called **buckets**, within which you can store and retrieve byte strings called **objects**.

The Amazon web interface allows you to interact with S3 simply by pointing and clicking. For example, figure 1.5 on the following page shows it being used to create a new bucket called `cloud4sciencebucket`, located within the US Standard region. (Amazon, like other cloud providers, operates many data centers around the world. The US Standard region is located in northern Virginia.) Such intuitive interfaces that can be used without any programming or even local software installation (beyond a web browser) on the part of the user are part of the attraction of cloud services.

S3 also defines a REST API that you can use to manipulate buckets and objects programmatically. Thus, instead of using the Amazon web interface, we could have created the bucket named `cloud4sciencebucket` via a PUT request on the URI `cloud4sciencebucket.s3.amazonaws.com`. The following shows the syntax of this PUT request, although omitting some of the header fields for simplicity.

```
PUT / HTTP/1.1
Host: cloud4sciencebucket.s3.amazonaws.com
Content-Length: length
Date: date
Authorization: authorization string
<CreateBucketConfiguration
            xmlns="http://s3.amazonaws.com/doc/2006-03-01/">
  <LocationConstraint>US Standard</LocationConstraint>
</CreateBucketConfiguration>
```

Similarly, a DELETE operation on the same URI requests deletion of the bucket that we just created, and a GET operation on that URI returns some or all of the objects that may have subsequently been placed in the bucket.

In later chapters, we describe such cloud service APIs and SDKs for a range of cloud infrastructure and platform services. Not covered in this book, but also interesting, are the APIs and SDKs provided by many of the SaaS offerings listed above: for example, Dropbox, Google Docs, LinkedIn, Science Exchange, and GitHub. (Not all SaaS provide APIs: Google Scholar and ResearchGate, sadly, do not.) The fact that we can easily access most cloud services both via web browser and programmatically is one of the reasons why cloud computing has proved so impactful.

Finally, we show how an SDK simplifies interactions with cloud services. The following Python code uses the Boto3 SDK to interact with Amazon S3. We obtain an S3 resource; delete the bucket created previously with the REST API; create the bucket again; and upload a file to the newly created bucket.

```python
import boto3
s3 = boto3.resource('s3')

# Delete the bucket previously created with the REST API
s3.Bucket('cloud3sciencebucket').delete()

# Create that bucket again, specifying location
bucket = s3.create_bucket(Bucket = 'cloud4sciencebucket',
                          CreateBucketConfiguration={
                          'LocationConstraint': 'us-standard'})

# Upload a file 'test.jpg' into the newly created bucket
bucket.put_object(Key='test.jpg', Body=open('test.jpg', 'rb'))
```

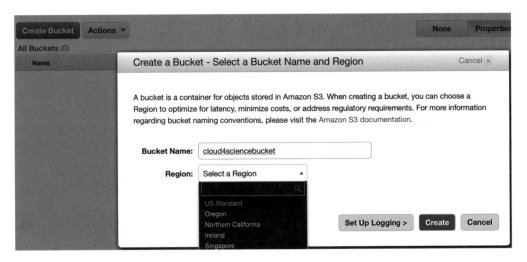

Figure 1.5: The Amazon S3 web interface at `console.aws.amazon.com/s3`, here being used to create a new bucket called `cloud4sciencebucket` in the US Standard region.

11

1.4.2 Local and Cloud-hosted Applications

The fact that we can, in a few lines, write programs that result in sophisticated actions occurring in a cloud service is exciting. But where should those programs run? One obvious location is your laptop or workstation, and indeed that may be the right place for many purposes. For example, we might use a slightly expanded version of the example program to upload 1,000 files from our laptop to S3.

However, in other cases, we want to run a program elsewhere: for example, because we want the program to keep running once we close our laptop, we cannot easily install required software on our local computer, or our program is intended to provide services to other people. In such cases, it is natural to run our program in the cloud. We discuss this topic in detail in part II of this book, where we see that we can create a cloud-hosted virtual computer, via either web interfaces or APIs/SDKs, much as we created a bucket in the example on the previous page.

In summary, the cloud can be viewed both as a source of services and as a place to run programs. Cloud services can be accessed from web browsers or from programs—programs that can themselves run locally or in the cloud. It is this diversity of usage modalities, and the relative simplicity of the methods by which these usage modalities are employed, that account for the power of the cloud.

1.5 Tools Used in This Book

We make extensive use in both this book and its supporting online notebooks of some standard tools that go beyond the world of cloud computing: the Python programming language, Jupyter web-based computing tool, GitHub version control and collaboration system, and Globus research data management service. We recommend that any researcher who aspires to become proficient in scientific computing master all four systems. All are quite accessible and are supported by excellent online resources. Time spent mastering them will be repaid many times over in more productive research. We give a brief introduction to each here.

1.5.1 Python

You need some basic programming knowledge to get the most out of this book. Most science and engineering students know the Python programming language, so we use Python for our programming examples. If you do not know Python, the book should still be interesting, but trust us, Python is easy and also tremendously fun. Learn the basics, at least.

The easiest way to get Python working on your computer is to install the free Anaconda distribution provided by Continuum Analytics `continuum.io/downloads`. This distribution includes multiple tools for installing and updating both Python and installed packages. It is separate from any OS-level version of Python, and is easy to completely uninstall. It works well on Windows, Mac, and Linux.

Alternatively, you can also create your Python environment manually, installing Python, package managers, and Python packages separately. Packages like NumPy and Pandas can be difficult to get working, however, particularly on Windows Anaconda simplifies this setup considerably, regardless of your OS.

1.5.2   Jupyter: An Interactive, Web-based Computing Tool

To facilitate access to the various methods and tools presented in this book, we provide complete source code for most code examples in the form of **Jupyter notebooks**. Jupyter Notebook, or simply **Jupyter**, is a web application that allows you to create and share documents ("notebooks") containing live code, equations, visualizations, and explanatory text. Figure 1.6 shows what Jupyter looks like in your web browser. The code for the notebook in this figure is in the code repository as notebook 1, as documented in chapter 17.

To install Juypter for Python, use the Python package installer `pip` or download and install Anaconda from Continuum Analytics. Later in this book we demonstrate how to install Python and Jupyter as a virtual machine or Docker container running in a remote cloud server.

Our use of Jupyter emphasizes that cloud computing lends itself to interactive exploration. We make almost all of the examples in this book available as Jupyter notebooks. Most were developed by the authors during interactive sessions using one or more of the cloud platforms described in this book.

1.5.3   The GitHub Version Control System

We also recommend that you master **GitHub**. A version control system is a tool for keeping track of changes that have been made to a document over time. GitHub is a hosting service for projects that use the Git version control system. Both the Git tool and the GitHub site are increasingly often used by researchers, to create digital lab notebooks that record the data files, programs, papers, and other resources associated with a project, with automatic tracking of the changes that are made to those resources over time [240]. GitHub also makes it easy for collaborators to work together on a project, whether a program or a paper: changes made by each

This is a demo of Jupyter and its graphics capabilities.

This cell is a markdown cell

```
[1]: import numpy as np
     from matplotlib import pyplot
     import matplotlib.pyplot as plt
     from mpl_toolkits.mplot3d.axes3d import Axes3D
     from pylab import *
     %matplotlib inline
```

```
[2]: def f(x, y):
         return 1 - cos(y)*cos(x)
     y_vec = linspace(0, 2*pi, 100)
     x_vec = linspace(0, 2*pi, 100)
     X,Y = meshgrid(x_vec, y_vec)
     Z = f(X, Y).T
```

```
[4]: fig = plt.figure(figsize=(9,6))
     ax = fig.add_subplot(1, 1, 1, projection='3d')
     p = ax.plot_surface(X, Y, Z, rstride=1, cstride=1,
                         cmap=cm.coolwarm, linewidth=0,
                         antialiased=False)
     cb = fig.colorbar(p, shrink=0.5)
```

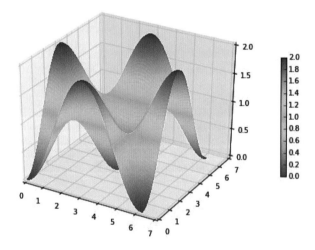

Figure 1.6: A sample Jupyter notebook. This notebook includes four cells. The first is a markdown cell, i.e., one containing text. The following three provide Python code that can be run from your web browser, with the last producing a visualization.

contributor are recorded and can easily be reconciled. For example, we used GitHub to create this book, with both the authors and reviewers checking in changes and comments at different times and time zones. Ram [222] provides a nice description of how Git/GitHub can be used to promote reproducibility and transparency in research. We also use GitHub to provide access to the online notebooks that accompany this text. You can access these notebooks at `Cloud4SciEng.org`.

1.5.4 Globus

Globus research data management service

We also use the Globus software as a service `globus.org` in our examples, and think that you will find it useful as well. This cloud-hosted service implements research data, identity, and credential management capabilities that can greatly simplify the development of cloud applications that need to access resources located on university campuses, national computing centers, and other facilities. It comprises a set of cloud-hosted software-as-a-serviceGlobus research data management service!as software as a service services (data transfer and synchronization, authorization, data sharing, data publication, data search, and groups), plus a simple software component, Globus Connect, deployable on computers associated with storage systems, including laptops, lab servers, campus compute clusters, cloud storage, and scientific instruments. REST APIs and a Python SDK simplify integration into applications. We provide more information about Globus in later chapters, as we introduce examples that demonstrate its use to manage access to data, replicate data across sites, and publish data, among other things.

1.6 Summary

Pioneering scientists and engineers are already using the cloud in their work, often in areas in which new data sources or modeling methods require more or different resources than are easily available in the laboratory: for example, in the analysis of urban [76, 84, 266] and environmental [110, 117, 137, 142, 252] data and in biomedical data analysis and modeling [33, 73, 188, 203, 260]. We are certain that many more researchers are reaching similar turning points and are thus finding themselves needing to think about computing in new ways. This book is for them and for a generation of computer scientists and engineers who recognize that cloud computing is going to be essential for their careers.

It is almost a futile exercise to write a practical, hands-on book about a technology that is evolving as rapidly as cloud. Our choices of major vendors and services may well be rendered obsolete by new developments. Nevertheless, we expect that most core concepts and tools will remain valid for a long time. The

Unix operating system, hot in the 1970s, lives on in Linux today; the Python programming language has been with us for 25 years. As new and better ideas emerge with implications for science and engineering, whether incremental or revolutionary, we will strive to update the online resources and, as and when feasible, produce revised versions of this text.

1.7 Resources

The U.S. National Institutes of Standards and Technology provides a useful definition of cloud computing, as "a model for enabling ubiquitous, convenient, on-demand network access to a shared pool of configurable computing resources (e.g., networks, servers, storage, applications and services) that can be rapidly provisioned and released with minimal management effort or service provider interaction" [197].

We recommend Charles Severance's *Python for Informatics: Exploring Information* [233], which covers basic Python and provides material relevant to web data and MySQL. This book is freely available online and is supported by excellent online lectures and exercises.

There are many Jupyter resources: see, for example, the main Jupyter site `jupyter.org`. Fernando Pérez and Brian Granger have an excellent blog called the "State of Jupyter" [219], that is both history and a look ahead.

Each public cloud has a portal where you can learn about their services, obtain free accounts with modest-sized allocations for experimentation, and track your data resources and compute activities: Amazon `aws.amazon.com`, Microsoft `azure.microsoft.com`, and Google `cloud.google.com`. Amazon and Microsoft grant programs can provide larger allocations: see `aws.amazon.com/grants` and `research.microsoft.com/azure`. The latter sites list examples of cloud in research, as do reports by Gannon et al. [135] and Lifka et al. [182].

To access the NSF-funded Jetstream cloud you need an allocation through the XSEDE program. If you are a U.S. academic researcher you can qualify for an allocation. Details are at `jetstream-cloud.org` and `xsede.org`.

Part I

Managing Data in the Cloud

Building your own cloud
What you need to know
Using Eucalyptus
Part IV Using OpenStack

Security and other topics
Securing services and data
Solutions
History, critiques, futures **Part V**

Part III **The cloud as platform**

Data analytics	Streaming data	Machine learning	Research data portals
Spark & Hadoop	Kafka, Spark, Beam	Scikit-Learn, CNTK,	DMZs and DTNs, Globus
Public cloud Tools	Kinesis, Azure Events	Tensorflow, AWS ML	Science gateways

Managing data in the cloud
File systems
Object stores
Databases (SQL)
NoSQL and graphs
Warehouses
Part I Globus file services

Computing in the cloud
Virtual machines
Containers – Docker
MapReduce – Yarn and Spark
HPC clusters in the cloud
Mesos, Swarm, Kubernetes
Part II HTCondor

Part I:
Managing Data in the Cloud

Data storage was the first manifestation of the cloud. Amazon's S3 public data storage service was launched in 2006. In 2008, Dropbox was introduced as a cloud service that could replace having to share files by passing around USB flash drives. That same year, Microsoft introduced its SkyDrive cloud storage service, later integrated with a service called Live Mesh that allowed synchronization across multiple machines, and in 2014 rebranded as OneDrive due to a lawsuit over the use of the word "Sky." Google introduced Google Drive in 2012.

These services all demonstrate the utility of a cloud service that allows you access to data anywhere, at any time, and on any device. However, they only represent one data storage model that is important for cloud computing. In this first part of the book we explore the following models.

- **File system** storage is the well-known model of organizing data into folders and directories. In the cloud, file storage is usually accessed by attaching a virtual disk to a virtual machine.

- **Blob storage**, where Blob is shorthand for *Binary Large Object*, provides a flat object model for data. It is extremely scalable, in ways that are challenging for file systems.

- **Databases** provide highly structured data collections. We consider three primary types of database in this book:

  1. Relational databases, which have a formal algebra of composition that can be invoked by the structured query language, SQL.

  2. Tables and NoSQL databases, which are more easily distributed over multiple machines.

3. Graph databases, in which data are represented as a graph of nodes and edges.

- **Data warehouses** that can support and enable search over massive amounts of data.

The two chapters in part I first explore these different models and then describe the landscape of storage offerings from the various cloud providers. One important capability of cloud data models for science is the support that they provide for managing data remotely; each provides an API and SDK that can be used to script data management tasks. We use various simple examples to illustrate the use of Python SDKs for Amazon, Azure, Google, and OpenStack. We also describe the Globus file management services, which are of particular importance for scientific applications in which data are often produced and consumed outside the cloud and need to move seamlessly among different locations.

.

Chapter 2

Storage as a Service

*"As a general rule the most successful man in life is the man who has
the best information."*

—Benjamin Disraeli

Science is concerned above all with data: with their acquisition, preservation, organization, analysis, and exchange. Thus we begin this book with a discussion of major cloud data storage services. These services collectively support a wide range of data storage models, from unstructured objects to relational tables, and offer a variety of performance, reliability, and cost characteristics. Collectively they provide the scientist or engineer with a wonderfully rich, if initially daunting, set of data storage capabilities.

In this chapter and the next, we introduce important cloud data storage concepts and illustrate how these concepts are realized in major cloud storage systems, using a range of examples to show how to use these systems to outsource simple data management tasks. In later chapters, we build on this foundation to show how cloud storage systems can be used in conjunction with other cloud services to construct powerful data management and analysis capabilities, for example when a data store is combined with an event notification service and compute service to enable analysis of streaming data.

2.1 Three Motivating Examples

Many cloud services that we use routinely—services such as Box, Dropbox, OneDrive, Google Docs, YouTube, Facebook, and Netflix—are, above all, data services. Each runs in the cloud, hosts digital content in the cloud, and provides specialized methods for accessing, storing, and sharing that content. Each is built on one or more—often multiple—storage services that have been variously optimized for speed, scale, reliability, and/or consistency. Our interest here is in how the basic infrastructure components on which these applications are built can be applied to science and engineering problems. To address this question, we need to understand the various types of storage systems that are common in the cloud so that we may evaluate their relative merits. To provide context, we consider the following three science and engineering use cases.

UC1 A climate science laboratory has assembled a set of simulation output files, each in Network Common Data Form (NetCDF) format: some 20 TB in total. These data are to be made accessible via interactive tools running in a web portal. The data sizes are such that data need to be partitioned to enable distributed analyses over multiple machines running in parallel.

UC2 A seismic observatory is acquiring records describing experimental observations, each specifying the time of the observation, experimental parameters, and the measurement itself, in CSV format. There may be a total of 1,000,000 records totaling some 100 TB when they finish collecting. They need to store these data to enable easy access by a large team, and to permit tracking of the data inventory and its accesses.

UC3 A team of scientists operates a collection of several thousand instruments, each of which generates a data record every few seconds. The individual records are not large, but managing the aggregate stream of all outputs in order to perform analyses across the entire collection every few hours introduces data management challenges. This problem is similar to that of analyzing large web traffic or social media streams.

Each use case requires different storage and processing models. In the paragraphs and chapters that follow, we divide these scenarios into more specific data collection examples and show how each can map to specific cloud storage services.

2.2 Storage Models

Before reviewing specific cloud storage services, we say a few words about storage models: that is, the different ways in which data can be organized in a storage system. An exciting feature of cloud storage systems is that they support a wide range of different storage models: not just the file systems that most researchers use on a daily basis, but also more specialized models such as object stores, relational databases, table stores, NoSQL databases, graph databases, data warehouses, and archival storage. Furthermore, their implementations of these models are often highly scalable, adapting easily from megabytes to hundreds of terabytes and beyond, while avoiding the need for dedicated operations expertise on the part of the user. Given the many challenges faced by scientists and engineers as they struggle with rapidly growing data, cloud storage can thus be the answer to their prayers. But one needs to understand the properties of these different storage models in order to choose the right system.

The right storage model for a data collection can depend on not only the nature and size of the data, but also the analyses to be performed, sharing plans, and update frequencies, among other factors. We review here some of the more important storage models supported by cloud storage services and, for each, their principal capabilities and their pros and cons for various purposes. This material provides background for the detailed service descriptions in the next chapter.

2.2.1 File Systems

The one storage model with which every scientist and engineer is surely familiar is the file system, organized around a tree of directories or folders. This model has proven to be an extremely intuitive and useful data storage abstraction. The standard API for the Unix-derived version of the file system is called the **Portable Operating System Interface** (POSIX). We are all familiar with the POSIX file system: we use it every day on our Apple, Linux, or Windows computer. Using command line tools, graphical user interfaces, or APIs, we create, read, write, and delete *files* located within *directories*.

The file system storage model has important advantages. It allows for the direct use of many existing programs without modification: we can navigate a file system with familiar file system browsers, run programs written in our favorite analysis tool (Python, R, Stata, SPSS, Mathematica, etc.) on files, and share files via email. The file system model also provides a straightforward mechanism for representing hierarchical relationships among data—directories and files—and supports concurrent access by multiple readers. In the early 1990s, the POSIX model was extended to distributed network file systems and there were some attempts at wide area versions. By the late 1990s, the Linux Cluster community created **Lustre**, a true parallel file system that supports the POSIX standard.

The file system model also has disadvantages as a basis for science and engineering, particularly as data volumes grow. From a data modeling perspective, it provides no support for enforcing conventions concerning the representation of data elements and their relationships. Thus, while one may, for example, choose to store environmental data in NetCDF files, genomes in FASTA files, and experimental observations in comma-separated-value (CSV) files, the file system does nothing to prevent the use of inconsistent representations within those files. Furthermore, the rigid hierarchical organization enforced by a file system often does not match the relationships that one wants to capture in science. Lacking any information on the data model, file systems cannot help users navigate complex data collections. The file system model also has problems from a scalability perspective: the need to maintain consistency as multiple processes read and write a file system can lead to bottlenecks in file system implementations.

For these reasons, cloud storage services designed for large quantities of data frequently adopt different storage models, as we discuss next.

2.2.2   Object Stores

The object storage model, like the file system model, stores unstructured binary objects. In the database world, objects are often referred to as **blobs**, for *binary large object*, and we use that name here when it is consistent with terminology adopted by cloud vendors. An object/blob store simplifies the file system model in important ways: in particular, it eliminates hierarchy and forbids updates to objects once created. Different object storage services differ in their details, but in general they support a two-level folder-file hierarchy that allows for the creation of object **containers**, each of which can hold zero or more **objects**. Each object is identified by a unique identifier and can have various metadata associated with it. Objects cannot be modified once uploaded: they can only be deleted—or, in object stores that support versioning, replaced.

We can use this storage model to store the NetCDF data in use case UC1. As shown in figure 2.1 on the following page, we create a single container and store each NetCDF file in that container as an object, with the NetCDF filename as the object identifier. Any authorized individual who possesses that object identifier can then access the data via simple HTTP requests or API calls.

The object store model has important advantages in terms of simplicity, performance, and reliability. The fact that objects cannot be modified once created makes it easy to build highly scalable and reliable implementations. For example, each object can be replicated across multiple physical storage devices to increase resilience and (when there are many concurrent readers) performance, without any specialized synchronization logic in the implementation to deal with concurrent updates. Objects can be moved manually or automatically among storage classes with different performance and cost parameters.

The object store model also has limitations. It provides little support for organizing data and no support for search: A user must know an object's identifier in order to access it. Thus, an object store would likely be inadequate as a basis for organizing the 1,000,000 environmental records of UC2: We would need to create a separate index to map from file characteristics to object identifiers. Nor does an object store provide any mechanism for working with structured data. Thus, for example, while we could load each UC2 dataset into a separate object, a user would likely have to download the entire object to a computer to compute on its contents. Finally, an object store cannot easily be mounted as a file system or accessed with existing tools in the ways that a file system can.

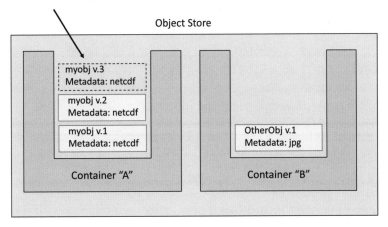

Figure 2.1: Object storage model with versioning. Each NetCDF file is stored in a separate container, and all versions of the same NetCDF file are stored in the same container.

2.2.3 Relational Databases

A **database** is a structured collection of data about entities and their relationships. It models real-world objects—both entities (e.g., microscopes, experiments, and samples, as in UC2 above) and relationships (e.g., "sample1" was tested on "July 24")—and captures structure in ways that allow these entities and relationships to be queried for analysis. A **database management system** (DBMS) is a software suite designed to safely store and efficiently manage databases and to assist with the maintenance and discovery of the relationships that databases represent. In general, a DBMS encompasses three components: its **data model** (which defines how data are represented), its **query language** (which defines how the user interacts with the data), and support for **transactions and crash recovery** (to ensure reliable execution despite system failures).

For a variety of reasons, science and engineering data often belong in a database rather than in files or objects. The use of a DBMS simplifies data management and manipulation and provides for efficient querying and analysis, durable and reliable storage, scaling to large data sizes, validation of data formats, and management of concurrent accesses.

While DBMSs in general, and cloud-based DBMSs in particular, support a wide variety of data formats, two major classes can be distinguished: relational and NoSQL. (We also discuss another class, graph databases, below.)

Relational DBMSs allow for the efficient storage, organization, and analysis of large quantities of tabular data: data organized as tables, in which rows represent entities (e.g., experiments) and columns represent attributes of those entities (e.g., experimenter, sample, result). The associated Structured Query Language (SQL) can then be used to specify a wide range of operations on such tables, such as compositions and joins. For example, the following SQL joins two tables, `Experiments` and `People`, to find all experiments performed by Smith:

```
select experiment-id from Experiments, People
where  Experiments.person-id = People.person-id
       and People.name = "Smith";
```

SQL statements can be executed with high efficiency thanks to sophisticated indexing and query planning techniques. Thus this join can be executed quickly even if there are millions of records in the tables being joined.

Many open source, commercial, and cloud-hosted relational DBMSs exist. Among the open source DBMSs, **MySQL** and **PostgreSQL** (often simply Postgres) are particularly widely used. Both MySQL and Postgres are available in cloud-hosted forms. In addition, cloud vendors offer specialized relational DBMSs that are designed to scale to particularly large data sizes.

Relational databases have two important properties. First, they support a relational algebra that provides a clear, mathematical meaning to the SQL language, facilitating efficient and correct implementations. Second, they support **ACID semantics**, a term that captures four important database properties: **A**tomicity (the entire transaction succeeds or fails), **C**onsistency (the data collection is never left in an invalid or conflicting state), **I**solation (concurrent transactions cannot interfere with each other), and **D**urability (once a transaction completes, system failures cannot invalidate the result).

2.2.4 NoSQL Databases

While relational DBMSs have long dominated the database world, other technologies have become popular for some application classes. A relational DBMS is almost certainly the right technology to use for highly structured datasets of moderate size. But if your data are less regular (if, for example, you are dealing with large amounts of text or if different items have different properties) or extremely large, you may want to consider a NoSQL DBMS. The design of these systems has typically been motivated by a desire to scale the quantities of data and number of users that can be supported, and to deal with unstructured data that are not easily represented in tabular form. For example, a **key-value store** can organize large numbers of records, each of which associates an arbitrary key with an arbitrary value. (A variant called a **document store** permits text search on the stored values.)

NoSQL databases have limitations relative to relational databases. The name NoSQL is derived from "non SQL," meaning that they do not support the full relational algebra. For example, they typically do not support queries that join two tables, such as those shown above.

Another definition of NoSQL is "not only SQL," meaning that most of SQL is supported but other properties are available. For example, a NoSQL database may allow for the rapid ingest of large quantities of unstructured data, such as the instrument events of UC3. Arbitrary data can be stored without modifications to a database schema, and new columns introduced over time as data and/or understanding evolves. NoSQL databases in the cloud are often distributed over multiple servers and also replicated over different data centers. Hence they often fail to satisfy all of the ACID properties. Consistency is often replaced by **eventual consistency**, meaning that database state may be momentarily inconsistent across replicas. This relaxation of ACID properties is acceptable if your concern is to respond rapidly to queries about the current state of a store's inventory. It may be unacceptable if the data in question are, for example, medical records.

Challenges of scale: The CAP theorem (from Foster et al. [125]). For many years, and still today, the big relational database vendors (Oracle, IBM, Sybase, Microsoft) were the mainstay of how data were stored. During the Internet boom, startups looking for low-cost alternatives to commercial relational DBMSs turned to MySQL and PostgreSQL. However, these systems proved inadequate for big sites as they could not cope well with large traffic spikes, as for example when many customers all suddenly wanted to order the same item. That is, they did not *scale*.

An obvious solution to scaling databases is to distribute and/or replicate data across multiple computers, for example by distributing different tables, or different rows from the same table. However, distribution and replication also introduce challenges, as we now explain. Let us first define some terms. In a system that comprises multiple computers:

- **Consistency** indicates that all computers see the same data at the same time.

- **Availability** indicates that every request receives a response about whether it succeeded or failed.

- **Partition tolerance** indicates that the system continues to operate even if a network failure prevents computers from communicating.

An important result in distributed systems (the "CAP Theorem" [78]) observes that it is not possible to create a distributed system with all three properties. This situation creates a challenge with large transactional datasets. Distribution is needed for high performance, but as the number of computers grows, so too does the likelihood of network disruption among computers [65]. As strict consistency cannot be achieved at the same time as availability and partition-tolerance, the DBMS designer must choose between high consistency or high availability for a particular system.

The right combination of availability and consistency will depend on the needs of the service. For example, in an e-commerce setting, we may choose high availability for a checkout process to ensure that revenue-producing requests to add items to a shopping cart are honored. Errors can be hidden from the customer and sorted out later. However, for order submission—when a customer submits an order—we should favor consistency, because several services (credit card processing, shipping and handling, reporting) need to access the data simultaneously.

2.2.5 Graph Databases

A graph is a data structure in which **edges** connect **nodes**. Graphs are useful when we need to search data based on relationships among data items. For example, in UC2, measurements from different experiments might be related by their use of the same measurement modality; in a database of scientific publications, publications can be represented as nodes and citations, shared authors, or even shared concepts as edges. Often graph databases are built on top of existing NoSQL databases.

2.2.6 Data Warehouses

The term data warehouse is commonly used to refer to data management systems optimized to support analytic queries that involve reading large datasets. Data warehouses have different design goals and properties than do DBMSs. For example, a medical center's clinical DBMS is typically designed to enable many concurrent requests to read and update information on individual patients (e.g., "what is Ms. Smith's current weight?" or "Mr. Jones was prescribed Aspirin"). Data from this DBMS are uploaded periodically (e.g., once a day) into the medical center's data warehouse to support aggregate queries such as "What factors are correlated with length of stay?" As we discuss in the next section, several cloud vendors offer data warehouse solutions that can scale to extremely large data volumes.

2.3 The Cloud Storage Landscape

The major public cloud companies provide a rich collection of storage services. The cloud providers described here are Amazon Web Services (hereafter referred to as **Amazon**), Microsoft Azure (hereafter **Azure**), and Google Cloud (hereafter, **Google**). Table 2.1 lists selected offerings from these three major vendors. (When a single vendor has multiple services in a category, those services tend to have quite different characteristics.) There are also other, more specialized storage services not listed in the table, some of which we mention in the following. We expand upon each of the rows in this table in the text that follows.

We do not list OpenStack options in the table because they are dependent upon the specific deployment, are not part of the OpenStack standard, and are not as extensive as those provided by the three main public clouds. Nevertheless, some standards for file services exist, as we discuss in section 2.3.7 on page 34.

2.3.1 File Systems

File systems (also referred to as **file shares**) are virtual data drives that can be attached to virtual machines. We describe the following services in greater detail in part II of the book. Amazon's **Elastic Block Store** (EBS) and **Elastic File System** (EFS) services offer related but different services. EBS is a device that

Table 2.1: Storage as a service options from major public cloud vendors.

Model	Amazon	Google	Azure
Files	Elastic File System (EFS), Elastic Block Store (EBS)	Google Cloud attached file system	Azure File Storage
Objects	Simple Storage Service (S3)	Cloud Storage	Blob Storage Service
Relational	Relational Data Service (RDS), Aurora	Cloud SQL, Spanner	Azure SQL
NoSQL	DynamoDB, HBase	Cloud Datastore, Bigtable	Azure Tables, HBase
Graph	Titan	Cayley	Graph Engine
Warehouse analytics	Redshift	BigQuery	Data Lake

you can mount onto a single Amazon EC2 compute server instance at a time; it is designed for applications that require low-latency access to data from a single EC2 instance. For example, you might use it to store working data that are to be read and written frequently by an application, but that are too large to fit in memory. EFS, in contrast, is a general-purpose file storage service. It provides a file system interface, file system access semantics (e.g., strong consistency, file locking), and concurrently-accessible storage for many Amazon EC2 instances. You might use EFS to hold state that is to be read and written by many concurrent processes. Note that both EBS and EFS can be accessed directly only by EC2 instances, that is, from inside the Amazon cloud.

Google Compute Engine has a different attached storage model. There are three types of attached disks (and also a way to attach an object store). The cheapest, **persistent disks**, can be up to 64 TB in size. **Local SSD** (solid state disk) is higher performance but more expensive and can be up to 3 TB. Finally, **RAM disk** is in-memory, limited to 208 GB, and expensive. Persistent disks can be accessed anywhere in a zone, but SSD and RAM are only accessible by the instance to which they are attached.

The **Azure File Storage** service allows users to create file shares in the cloud that can be accessed by a special protocol, **SMB**, that allows Microsoft Windows VMs and Linux VMs to mount these file shares as standard parts of their file system. These file shares can also be mounted on your Windows or Mac.

2.3.2   Object Stores

Amazon's **Simple Storage Service** (S3) was historically its first cloud service. It is highly popular: as of 2016, it reportedly holds trillions of objects in billions of containers, which S3 calls **buckets**. S3 is a classic object store, with all of the properties listed in section 2.2.2 on page 24. We describe S3 in more detail in section 3.2 on page 38, where we also present examples of its use. The related **Glacier** service is designed for long-term, secure, durable, extremely low cost data archiving. Access times for an object in Glacier may be several hours, so this is not for applications that need rapid data access.

Google's **Cloud Storage**, like Amazon S3, provides a basic object storage system that is durable, replicated and highly available. It supports three storage tiers, each with different performance and price levels. The most expensive is **Standard** multiregional, the mid-range tier is **Regional** (DRA), and the bottom tier is **Nearline**. Standard storage is for data you expect to access often, DRA is for batch jobs for which response time is not a critical issue, and Nearline is for cold storage and disaster recovery. Google Cloud also has **Coldline**, which is similar to AWS Glacier.

Azure Storage offers a suite of services with a similar scope in terms of models supported to those provided by Amazon and Google. Azure provides the user with a unified view of many of its storage types associated with their account, as shown in figure 2.2 on the next page. This integration means that you can use the Azure Storage Explorer tool `storageexplorer.com` to see and manage all of these storage products from your PC or Mac. While Azure storage services were originally optimized for close integration with Microsoft Windows environments, Linux is now an important part of Azure, so this differentiation is less noticeable.

The Azure **Blob** storage service, like Amazon's S3, is concerned with highly reliable storage of unstructured objects, which Microsoft calls *blobs*. Like Amazon and Google Cloud, Azure blob storage has tiered storage and pricing. The tiers are *hot* for frequently accessed data and *cool* for data that are accessed less often.

2.3.3   NoSQL Services

Amazon's **DynamoDB** is a powerful NoSQL database based on an extensible key-value model: for each row, the primary key column is the only required attribute, but any number of additional columns can be defined, indexed, and made searchable in various ways, including full-text search via Elasticsearch. DynamoDB's rich feature set defies a concise description, but we illustrate some of its uses in section 3.2 on page 38. Related to this is Amazon **Elastic MapReduce** (EMR),

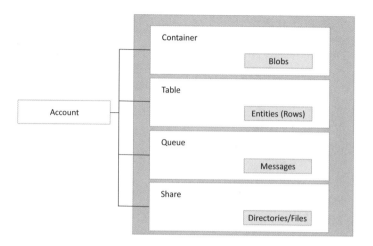

Figure 2.2: Major Azure storage types.

discussed in section 8.3 on page 143, which allows analysis of large quantities of data with Spark and other data analytics platforms.

Cloud Bigtable, Google's highly scalable NoSQL database service, is the same database that powers many core Google services, including Search, Analytics, Maps, and Gmail. Bigtable maps two arbitrary strings (row key and column key) and a timestamp (permitting versioning and garbage collection) to an associated arbitrary byte array. It is designed to handle such large and sparse datasets in a manner that is efficient in space used and that supports massive workloads, while providing low latency and high bandwidth. You deploy Bigtable on a Google-hosted cluster, which can be dynamically resized if needed. The open source Apache HBase database system, fundamental to the Apache Hadoop system, is compatible with Bigtable. Google's **Cloud Datastore** has many similarities to Bigtable. An important difference is that it implements ACID semantics and thus the user does not have to wait for possible inconsistencies to be resolved, as is required for Bigtable. Cloud Datastore has a much richer set of SQL-like operators than do many of the other NoSQL systems described here.

The Azure **Table** storage service is a simple NoSQL key-value store, designed to support the highly reliable storage of any large number of key-value pairs. It is similar to Amazon DynamoDB. Its query capabilities are limited, but it can support many queries at modest cost. Azure **HDInsight** provides an implementation of the Hadoop storage service hosted on Azure cloud computers, with implementations of popular big data tools, including Spark, the HBase NoSQL database, and the Hive SQL database implemented to run efficiently and scalably on top of that Hadoop fabric. We describe this service in more detail in part III. **DocumentDB** is a NoSQL service, like Table, but supporting full text indexing and query, albeit at higher cost due to the greater resources needed for indexing.

2.3.4 Relational Databases

Relational databases are a mature technology, so the main innovation in the cloud is deployments that scale to especially large sizes.

Amazon's **Relational Database Service** (RDS) allows you to set up a conventional relational database (e.g., MySQL or Postgres) on Amazon computers, thus permitting MySQL and Postgres applications to be ported to Amazon without change. The MySQL-compatible Amazon **Aurora** service provides higher scalability, performance, and resilience than an RDS MySQL instance: it can scale to many terabytes, replicate data across data centers (known as availability zones), and create many read replicas to support large numbers of concurrent reads.

Google's **Cloud SQL** relational database service has similar capabilities to those provided by such services in Amazon and Azure. Their **Spanner** system [101] is a globally distributed relational database that provides ACID transactions and SQL semantics with high scaling and availability.

Azure's **SQL Database** provides a relational database service similar to Amazon RDS. It is based on their mature SQL Server technology and is highly available and scalable.

2.3.5 Warehouse Analytics

Cloud data warehouses are designed specifically for running analytics queries over large collections. You interact with them from the cloud portal or via REST APIs. We discuss these systems in greater detail in part III.

Amazon **Redshift** is a data warehouse system, designed to support high-performance execution of analytic and reporting workloads against large datasets.

For massive data analytics, Google provides the **BigQuery** petascale data warehouse. BigQuery is fully distributed and replicated, so durability is not an issue. It also supports SQL query semantics.

The Azure **Data Lake** is a full suite of data analytics tools built on the open-source YARN and WebHDFS platforms.

2.3.6   Graphs and More

Each of the three cloud providers provides not only graph databases but also other services, including messaging and stream services that, like warehouses, are important tools for stream and log analytics. Messaging services allow applications to send and receive messages using what are referred to as **publish/subscribe** semantics. They allow one application to wait on a queue for a message to arrive while other applications prepare the message and send it to the queue. This data capability is important for many cloud applications that need to distribute work tasks or to process streams of incoming events. We discuss messaging applications in chapters 7 through 9.

Amazon's **Titan** extension to DynamoDB supports graph databases. Other Amazon services not listed in table 2.1 include the **Simple Queue Service** (SQS: see section 9.3 on page 167) and the **Elasticsearch Service**, a cloud-hosted version of the Elasticsearch open-source search and analytics engine. Amazon **Kinesis**, also discussed in section 9.3, supports analysis of stream data.

Google supports the open source graph database **Cayley**. Its **Cloud Pub/Sub** service provides messaging in a similar manner to Amazon SQS.

The Azure graph database, **Graph Engine**, is a distributed, in-memory large graph processing system. The **Queue** service is the Azure pub/sub service, similar to Google Cloud Pub/Sub. The **Azure Event Hubs** service is similar to Amazon Kinesis. We return to these services in chapter 9.

2.3.7   OpenStack Storage Services and Jetstream

The OpenStack open source cloud software supports only a few standard storage services: object storage, block storage, and file system storage. However, many research groups and some large companies such as IBM are investing substantial effort to improving this situation.

The OpenStack object storage service is called **Swift** (not to be confused with the Swift parallel scripting language [259] or Apple's Swift language). Like Amazon S3 and the Azure Blob service, Swift implements a REST API that users can use to store, delete, manage permissions of, and associate metadata with immutable unstructured data *objects* located within *containers*. These objects are replicated across multiple storage servers for fault tolerance and performance reasons, and can be accessed from anywhere.

The OpenStack **Shared File Systems** service, like the Amazon EFS and Azure File service, implements a file system model in the cloud environment. Users interact with this service by mounting remote file systems, called *shares*, on their virtual machine instances. They can also create shares, configuring the file system protocol supported; manage access to shares; delete shares; and configure rate limits and quotas, among other things. Shares can be mounted on any number of client machines, using NFS, CIFS, GlusterFS, or HDFS drivers. Shares can be accessed only from virtual machine instances running on the OpenStack cloud.

Many OpenStack deployments use the **Ceph** distributed storage system to manage their storage. This open source system supports object storage with interfaces compatible with Amazon S3 and OpenStack Swift. It also supports block-level and file-level storage.

We use the U.S. National Science Foundation's **Jetstream** cloud for OpenStack examples. Operated as part of the **XSEDE** supercomputer project [245] `xsede.org`, Jetstream is more experimental than the large public clouds because it is designed to support new classes of interactive scientific computing. Jetstream runs the OpenStack object store, based on Ceph, which implements the Swift API. The primary user interaction with Jetstream is through a system known as **Atmosphere** built by the University of Arizona as part of the NSF IPlant collaborative. Atmosphere is designed to manage virtual machines, data, and visualization tools for communities of scientists, and provides a volume management system for mounting external volumes on VMs. We explore Atmosphere in greater detail in section 5.5 on page 82. Jetstream also operates the Globus identity, group, and file management services, which we describe in the next chapter.

2.4 Summary

As this chapter has illustrated, the data storage models used in the cloud are as varied as the types of data and the types of processing that scientists would want to use. Let us return to our use cases and see how they map to the data types.

The first use case, involving climate simulation output files in NetCDF format, is clearly a case for an Amazon S3, Google Cloud Storage, or Azure Blob Storage. Each blob can be up to 1 TB in size (5 TB for S3). As we show in chapter 3, each service provides simple APIs that can be used to access data. Another solution for moving the data to and from S3 is to use the Globus file transfer protocols, which have been optimized for managing big data objects: see section 3.6 on page 51.

The second use case, involving 1,000,000 records describing experimental observations, could also be handled with simple blob storage, but the cloud presents us with better solutions. The simplest is to use a standard relational SQL database, but the merits of this approach depend on how strict we are with the schema that describes the data. Do all records have the same structure, or do some have fields that others do not? In the latter case, a NoSQL database may be a superior solution. Other factors are scale and the possible need for parallel access. Cloud NoSQL stores like Azure Tables, Amazon DynamoDB, and Google Bigtable, are massively scalable and replicated. Unlike conventional SQL database solutions, they are designed for highly parallel massive data analysis.

The third use case, involving a massive set of instrument event records, is also appropriate for cloud NoSQL databases. However, data warehouses such as Amazon Redshift and Azure Data Lake are designed to be complete platforms for performing data analytics on massive data collections. If our instrument records are streaming in real time, we can use event streaming tools based on publish/subscribe semantics, as we discuss in chapter 9.

2.5 Resources

The storage capabilities of the public clouds are rapidly evolving and thus it is important to consult the relevant documentation, which is easily accessed from the cloud portals: `aws.amazon.com` for Amazon, `azure.microsoft.com` for Azure, and `cloud.google.com` for Google.

Troy Hunt uses the example of his "have I been pwned" site, which uses the Azure Table service to enable rapid searches against more than one billion compromised accounts, to illustrate some of the pros and cons of Azure's Table, DocumentDB, and SQL Database services [158].

Chapter 3

Using Cloud Storage Services

"Collecting data is only the first step toward wisdom, but sharing data is the first step toward community."

—Henry Louis Gates Jr.

We introduced in chapter 2 a set of important cloud storage concepts and a range of cloud provider services that implement these concepts in practice. While the services of different cloud providers are often similar in outline, they invariably differ in the details. Thus here we describe, and use examples to illustrate the use of the services used in three major public clouds (Amazon, Azure, Google) and in one major open source cloud system, OpenStack. And because your science will often involve data that exist outside the cloud, we also describe how you can use Globus to move data between the cloud and other data centers and to share data with collaborators.

3.1 Two Access Methods: Portals and APIs

As we discussed in section 1.4 on page 8, cloud providers make available two main methods for managing data and services: portals and REST APIs.

Each cloud provider's web portal typically allows you to accomplish anything that you want to do with a few mouse clicks. We provide several examples of such web portals to illustrate how they work.

While such portals are good for performing simple actions, they are not ideal for the repetitive tasks, such as managing hundreds of data objects, that scientists need to do on a daily basis. For such tasks, we need an interface to the cloud that we can program. Cloud providers make this possible by providing REST APIs that programmers can use to access their services programmatically. For programming convenience, you will usually access these APIs via software development kits (SDKs), which give programmers language-specific functions for interacting with cloud services. We discuss the Python SDKs here. The code below is all for Python 2.7, but is easily converted to Python 3.

Each cloud has special features that make it unique, and thus the different cloud provider's REST APIs and SDKs are not identical. Two efforts are under way to create a standard Python SDK: **CloudBridge** [11] and **Apache Libcloud** `libcloud.apache.org`. While both aim to support the standard tasks for all clouds, those tasks are only the lowest common denominator of cloud capabilities; many unique capabilities of each cloud are available only through the REST API and SDK for that platform. At the time of this writing, Libcloud is not complete; we will provide an online update when it is ready and fully documented. However, we do make use of CloudBridge in our OpenStack examples.

Building a data sample collection in the cloud. We use the following simple example throughout this chapter to illustrate the use of Amazon, Azure, and Google cloud storage services. We have a collection of data samples stored on our personal computer and for each sample we have four items of metadata: item number, creation date, experiment id, and a text string comment. To enable access to these samples by our collaborators, we want to upload them to cloud storage and to create a searchable table, also hosted in the cloud, containing the metadata and cloud storage URL for each object, as shown in figure 3.1 on the following page.

We assume that each data sample is in a binary file on our personal computer and that the associated metadata are contained in a comma separated value (CSV) file, with one line per item, also on our personal computer. Each line in this CSV file has the following format:

```
item id, experiment id, date, filename, comment string
```

3.2 Using Amazon Cloud Storage Services

Our Amazon solution to the example problem uses S3 to store the blobs and DynamoDB to store the table. We first need our Amazon key pair, i.e., access key plus secret key, which we can obtain from the Amazon **IAM Management Console**. Having created a new user, we select the *create access key* button to create our security credentials, which we can then download, as shown in figure 3.2 on the following page.

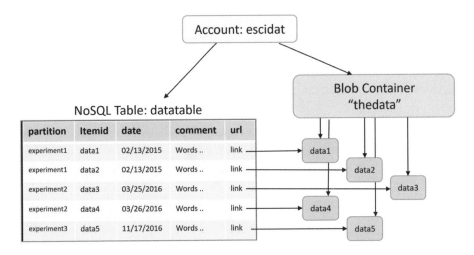

Figure 3.1: The simple cloud storage use case that we employ in this chapter involves the upload of a collection of data blobs to cloud storage and the creation of a NoSQL table containing metadata.

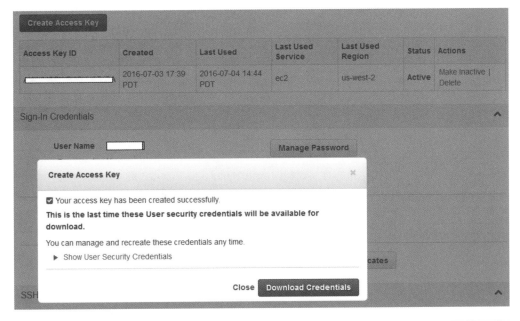

Figure 3.2: Downloading security credentials from the Amazon IAM Management Console.

We can proceed to create the required S3 bucket, upload our blobs to that bucket, and so forth, all from the Amazon web portal. (We showed in figure 1.5 on page 11 the use of this portal to create a bucket.) However, there are a lot of blobs, so we instead use the Amazon Python Boto3 SDK for these tasks. Details on how to install this SDK are found in the link provided in the Resources section at the end of this chapter.

Boto3 considers each service to be a **resource**. Thus, to use the S3 system, we need to create an S3 resource object. To do that, we need to specify the credentials that we obtained from the IAM Management Console. Several ways can be used to provide these credentials to our Python program. The simplest is to provide them as special named parameters to the resource instance creation function, as follows.

```
import boto3
s3 = boto3.resource('s3',
    aws_access_key_id='YOUR ACCESS KEY',
    aws_secret_access_key='your secret key' )
```

This approach is not recommended from a security perspective, since credentials in code have a habit of leaking, if, for example, code is pushed to a shared repository: see section 15.1 on page 315. Fortunately, this method is needed only if your Python program is running as a separate service on a machine or a container that does not have access to your security keys. If you are running on your own machine, the proper solution is to have a home directory .aws that contains two protected files: config, containing your default Amazon region, and credentials, containing your access and secret keys. If we have this directory in place, then the access key and secret key parameters are not needed.

Having created the S3 resource object, we can now create the S3 bucket, datacont, in which we will store our data objects. The following code performs this action. Note the (optional) second argument to the create_bucket call, which specifies the geographic region in which the bucket should be created. At the time of writing, Amazon operates in 13 regions; region us-west-2 is located in Oregon.

```
import boto3
s3 = boto3.resource('s3')
s3.create_bucket(Bucket='datacont', CreateBucketConfiguration={
    'LocationConstraint': 'us-west-2'})
```

Now that we have created the new bucket, we can load our data objects into it with a command such as the following.

```
# Upload a file, 'test.jpg' into the newly created bucket
s3.Object('datacont', 'test.jpg').put(
    Body=open('/home/mydata/test.jpg', 'rb'))
```

Having learned how to upload a file to S3, we can now create the DynamoDB table in which we will store metadata and references to S3 objects. We create this table by defining a special key that is composed of a **PartitionKey** and a **RowKey**. NoSQL systems such as DynamoDB are distributed over multiple storage devices, which enable constructing extremely large tables that can then be accessed in parallel by many servers, each accessing one storage device. Hence the table's aggregate bandwidth is multiplied by the number of storage devices. DynamoDB distributes data via row: for any row, every element in that row is mapped to the same device. Thus, to determine the device on which a data value is located, you need only look up the **PartitionKey**, which is hashed to an index that determines the physical storage device in which the row resides. The **RowKey** specifies that items are stored in order sorted by the **RowKey** value. While it is not necessary to have both keys, we also illustrate the use of **RowKey** here. We can use the following code to create the DynamoDB table.

```
dyndb = boto3.resource('dynamodb', region_name='us-west-2' )

# The first time that we define a table, we use
table = dyndb.create_table(
    TableName='DataTable',
    KeySchema=[
        { 'AttributeName': 'PartitionKey', 'KeyType': 'HASH'},
        { 'AttributeName': 'RowKey', 'KeyType': 'RANGE' }
    ],
    AttributeDefinitions=[
        { 'AttributeName': 'PartitionKey', 'AttributeType': 'S' },
        { 'AttributeName': 'RowKey',       'AttributeType': 'S' }
    ]
)

# Wait for the table to be created
table.meta.client.get_waiter('table_exists')
                        .wait(TableName='DataTable')

# If the table has been previously defined, use:
# table = dyndb.Table("DataTable")
```

We are now ready to read the metadata from the CSV file, move the data objects into the blob store, and enter the metadata row into the table. We do this as follows. Recall that our CSV file format has **item[3]** as filename, **item[0]** as itemID, `item[1]` as experimentID, `item[2]` as date, and `item[4]` as comment. Note that we need to state explicitly, via `ACL='public-read'`, that the URL for the data file is to be publicly readable. The complete code is in notebook 2.

```
import csv
urlbase = "https://s3-us-west-2.amazonaws.com/datacont/"
with open('\path-to-your-data\experiments.csv', 'rb') as csvfile:
    csvf = csv.reader(csvfile, delimiter=',', quotechar='|')
    for item in csvf:
        body = open('path-to-your-data\datafiles\\'+item[3], 'rb')
        s3.Object('datacont', item[3]).put(Body=body)
        md = s3.Object('datacont', item[3]).Acl()
            .put(ACL='public-read')
        url=urlbase +item[3]
        metadata_item={'PartitionKey': item[0], 'RowKey': item[1],
            'description' : item[4], 'date' : item[2], 'url':url}
        table.put_item(Item=metadata_item)
```

3.3   Using Microsoft Azure Storage Services

We first note some basic differences between your Amazon and Azure accounts. As we described above, your Amazon account ID is defined by a pair consisting of your access key and your secret key. Similarly, your Azure account is defined by your personal ID and a subscription ID. Your personal ID is probably your email address, so that is public; the subscription ID is something to keep secret.

We use Azure's standard blob storage and Table service to implement the example. The differences between Amazon DynamoDB and the Azure Table service are subtle. With the Azure Table service, each row has the fields `PartitionKey`, `RowKey`, `comments`, `date`, and `URL` as before, but this time the `RowKey` is a unique integer for each row. The `PartitionKey` is used as a hash to locate the row into a specific storage device, and the RowKey is a unique global identifier for the row.

In addition to these semantic differences between DynamoDB and Azure Tables, there are fundamental differences between the Amazon and Azure object storage services. In S3, you create buckets and then create blobs within a bucket. S3 also provides an illusion of folders, although these are actually just blob name prefixes (e.g., `folder1/`). In contrast, Azure storage is based on **Storage Accounts**, a higher level abstraction than buckets. You can create as many storage accounts as you want; each can contain five different types of objects: blobs, containers, file shares, tables, and queues. Blobs are stored in bucket-like containers that can also have a pseudo directory-like structure, similar to S3 buckets.

Given your user ID and subscription ID, you can use the Azure Python SDK to create storage accounts, much as we create buckets in S3. However, we find it easier to use the Azure portal. Login and click on (storage account) in the menu on the left to bring up a panel for storage accounts. To add a new account, click on the (+) sign at the top of the panel. You need to supply a name and some additional parameters such as location, duplication, and distribution. Figure 3.3 shows the storage account that we added, called `escistore`.

One big difference between S3 and Azure Storage accounts is that each storage account comes with two unique access keys, either of which can be used to access and modify the storage account. Unlike S3, you do not need the subscription ID or user ID to add containers, tables, blobs or queues; you only need a valid key. You can also invalidate either key and generate new keys at any time from the portal. The reason for having two keys is that you can use one key for your long-running services that use that storage account and the other to allow another entity temporary access. By regenerating that second key, you terminate access by the third party.

Azure storage accounts are, by default, private. You can also set up a public storage account, as we show in an example on the next page, and grant limited, temporary access to a private account by creating, from the portal, a **Storage Access Signature** for the account. Various access right properties can be configured in this signature, including the period for which it is valid.

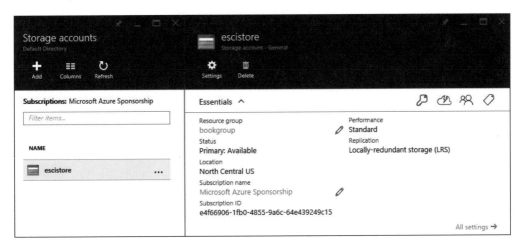

Figure 3.3: Azure portal after the storage account has been created.

Having created the storage account and installed the SDK (see section 3.8 on page 57), we can proceed to the initialization as follows. The `create_table()` function returns true if a new table was created and false if the table already exists.

```
import azure.storage
from azure.storage.table import TableService, Entity
from azure.storage.blob import BlockBlobService
from azure.storage.blob import PublicAccess
# First, access the blob service
block_blob_service = BlockBlobService(account_name='escistore',
    account_key='your storage key')
block_blob_service.create_container('datacont',
    public_access=PublicAccess.Container)
# Next, create the table in the same storage account
table_service = TableService(account_name='escistore',
                             account_key='your account key')
if table_service.create_table('DataTable'):
    print("Table created")
else:
    print("Table already there")
```

The code to upload the data blobs and to build the table is almost identical to that used for Amazon S3 and DynamoDB. The only differences are the lines that first manage the upload and then insert items into the table. To upload the data to the blob storage we use the `create_blob_from_path()` function, which takes three parameters: the container, the name to give the blob and the path to the source, as shown in the following.

```
import csv
with open('\path-to-your-data\experiments.csv', 'rb') as csvfile:
   csvf = csv.reader(csvfile, delimiter=',', quotechar='|')
   for item in csvf:
     print(item)
     block_blob_service.create_blob_from_path(
             'datacont', item[3],
             "\path-to-your-files\datafiles\\"+item[3]
             )
     url="https://escistore.blob.core.windows.net/datacont/"+item[3]
     metadata_item = {'PartitionKey':item[0], 'RowKey':item[1],
           'description' : item[4], 'date' : item[2], 'url':url}
     table_service.insert_entity('DataTable', metadata_item)
```

A nice desktop tool called Azure Storage Explorer is available for Windows, Macs, and Linux. We tested the code above with a single CSV file with only four lines and four small blobs. Figure 3.4 shows the Storage Explorer views of the blob container and table contents.

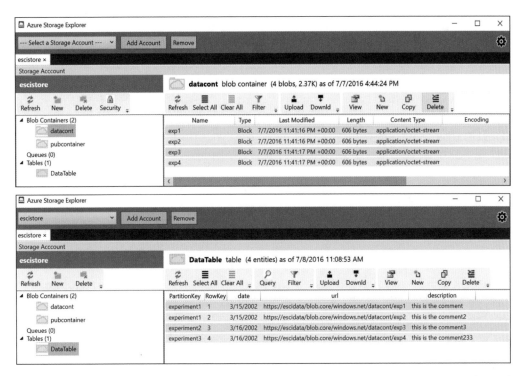

Figure 3.4: Azure Storage Explorer view of the contents of container `datacont` (above) and the table `DataTable` (below) in storage account `escidata`.

Queries in the Azure table service Python SDK are limited to searching for `PartitionKey`; however, simple filters and projections are possible. For example, you can search for all rows associated with `experiment1` and then select the `url` column as shown below.

```
tasks = table_service.query_entities('DataTable',
        filter="PartitionKey eq 'experiment1'", select='url')
for task in tasks:
    print(task.url)
```

Our trivial dataset has two data blobs associated with `experiment1`, and thus the query yields two results:

`https://escistore.blob.core.windows.net/datacont/exp1`

`https://escistore.blob.core.windows.net/datacont/exp2`

Similar queries are possible with Amazon DynamoDB. The Azure Python code for this example is in notebook 3.

3.4 Using Google Cloud Storage Services

The Google Cloud has long been a central part of Google's operations, but it is relatively new as a public platform to rival Amazon and Azure. Google's Google Docs, Gmail, and data storage services are well known, but their first foray into providing computing and data services of potential interest to science was **Google AppEngine**, which is not discussed here. Recently they have pulled many of their internally used services together into a public platform called **Google Cloud**, which includes their data storage services, their NoSQL services Cloud Datastore and Bigtable, and various computational services that we describe in part III. To use Google Cloud, you need an account. Google currently offers a small but free 60-day trial account. Once you have an account, you can install the Google Cloud SDK which consists of the `Google Cloud` command-line tool and the `gsutil` package. These tools are available for Linux, Mac OS, and Windows.

To get going, you need to install the `gsutil` package and then execute `gcloud init`, which prompts you to log into your Google account. You also need to create and/or select a project to work on. Once this is done, your desktop machine is authenticated and authorized to access the Google Cloud Platform services. While this is convenient, you must do a bit more work to write Python scripts that can access your resource from anywhere. We discuss this topic later.

For now, if we bring up our Jupyter notebook on our local machine, it is authenticated automatically. Creating a storage bucket and uploading data are now easy. You can create a bucket from the console or programmatically. Note that your bucket name must be unique across all of Google Cloud, so when creating a bucket programmatically, you may wish to use a universally unique identifier (UUID) as the name. For simplicity, we do not do that here.

```
from gcloud import storage
client = storage.Client()
# Create a bucket with name 'afirstbucket'
bucket = client.create_bucket('afirstbucket')
# Create a blob with name 'my-test-file.txt' and load some data
blob = bucket.blob('my-test-file.txt')
blob.upload_from_string('this is test content!')
blob.make_public()
```

The blob is now created and can be accessed at the following address.

`https://storage.googleapis.com/afirstbucket/my-test-file.txt`

Google Cloud has several NoSQL table storage services. We illustrate the use of two here: Bigtable and Datastore.

3.4.1 Google Bigtable

Bigtable is the progenitor of Apache HBase, the NoSQL store built on the Hadoop Distributed File System (HDFS). Bigtable and HBase are designed for large data collections that must be accessed quickly by major data analytics jobs. Provisioning a Bigtable instance requires provisioning a cluster of servers. This task is most easily performed from the console. Figure 3.5 illustrates the creation of an instance called `cloud-book-instance`, which we provision on a cluster of three nodes.

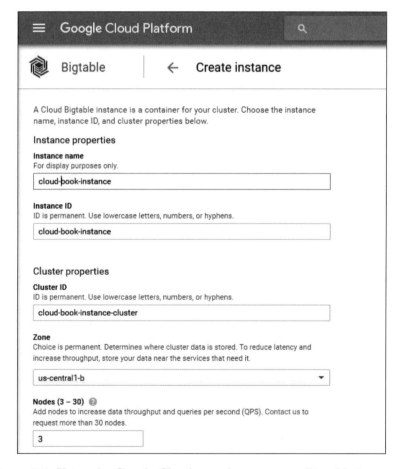

Figure 3.5: Using the Google Cloud console to create a Bigtable instance.

The following code illustrates how we can build a Bigtable instance and then create first a table and then a column family and a row in that table. Column families are the groups of columns that organize the contents of a row; each row has a single unique key.

```
from gcloud import bigtable
clientbt = bigtable.Client(admin=True)
clientbt.start()
instance = clientbt.instance('cloud-book-instance')
table = instance.table('book-table')
table.create()
# Table has been created
column_family = table.column_family('cf')
column_family.create()

#now insert a row with key 'key1' and columns 'experiment', 'date',
#'link'
row = table.row('key1')
row.set_cell('cf', 'experiment', 'exp1')
row.set_cell('cf', 'date', '6/6/16')
row.set_cell('cf', 'link', 'http://some_location')
row.commit()
```

Bigtable is important because it is used with other Google Cloud services for data analysis and is an ideal solution for truly massive data collections. The Python API for Bigtable is a bit awkward to use because the operations are all asynchronous remote procedure calls (RPCs): when a Python **create()** call returns, the object created may not yet be available. For those who want to experiment, we provide in notebook 4 code illustrating the use of Bigtable.

3.4.2 Google Cloud Datastore

Notebook 5 shows how to implement equivalent functionality by using Datastore. This service is replicated, supports ACID transactions and SQL-like queries, and is easier to use from the Python SDK than Bigtable. We create a table as follows.

```
from gcloud import datastore
clientds = datastore.Client()
key = clientds.key('blobtable')
```

To add an entity to the table, we write the following.

```
entity = datastore.Entity(key=key)
entity['experiment-name'] = 'experiment name'
entity['date'] = 'the date'
entity['description'] = 'the text describing the experiment'
entity['url'] = 'the url'
clientds.put(entity)
```

You can now implement the use case with ease. You first create a bucket `book-datacont` on the Google Cloud portal. Again, it is easiest to perform this operation on the portal because the name is unique across Google Cloud. (If you include a fixed name in your code, then the create call will fail when you rerun your program.) Furthermore, the bucket name `book-datacont` is now taken, so when you try this you need to choose your own unique name. You can then create a datastore table `book-table` as follows.

```
from gcloud import storage
from gcloud import datastore
import csv

client = storage.Client()
clientds = datastore.Client()
bucket = client.bucket('book-datacont')
key = clientds.key('book-table')
```

Your blob uploader and table builder can now be written as follows. The code is the same as that provided for the other systems, with the exception that we do not use a partition or row key.

```
with open('\path-to-your-data\experiments.csv', 'rb') as csvfile:
    csvf = csv.reader(csvfile, delimiter=',', quotechar='|')
    for item in csvf:
        print(item)
        blob = bucket.blob(item[3])
        data = open("\path-to-your-data\datafiles\\"+item[3], 'rb')
        blob.upload_from_file(data)
        blob.make_public()
        url = "https://storage.googleapis.com/book-datacont/"+item[3]
        entity = datastore.Entity(key=key)
        entity['experiment-name'] = item[0]
        entity['experiment-id'] = item[1]
        entity['date'] = item[2]
        entity['description'] = item[4]
        entity['url'] = url
        clientds.put(entity)
```

Datastore has an extensive query interface that can be used from the portal; some, but not all, features are also available from the Python API. For example, we can write the following to find the URLs for `experiment1`.

```
query = clientds.query(kind=u'book-table')
query.add_filter(u'experiment-name', '=', 'experiment1')
results = list(query.fetch())
urls = [result['url'] for result in results]
```

3.5 Using OpenStack Cloud Storage Services

OpenStack is used by IBM and Rackspace for their public cloud offering and also by many private clouds. We focus our OpenStack examples in this book on the NSF Jetstream cloud. As we discussed in chapter 2, Jetstream is not intended to duplicate the current public clouds, but rather to offer services that are tuned to the specific needs of the science community. One missing component is a standard NoSQL database service, so we cannot implement the data catalog example that we presented for the other clouds.

A Python SDK called CloudBridge works with Jetstream and other OpenStack-based clouds. (CloudBridge also works with Amazon but it is less comprehensive than Boto3.) To use CloudBridge, you first create a provider object, identifying the cloud with which you want to work and supplying your credentials, for example as follows.

```
from cloudbridge.cloud.factory import CloudProviderFactory, \
        ProviderList
js_config =
    {"os_username": "your user name",
        "os_password": "your password",
        "os_auth_url": "https://jblb.jetstream-cloud.org:35357/v3",
        "os_user_domain_name": "tacc",
        "os_tenant_name": "tenant name",
        "os_project_domain_name": "tacc",
         "os_project_name": "tenant name"
    }
js = CloudProviderFactory()\
        .create_provider(ProviderList.OPENSTACK, js_config)
```

You may now use the provider object reference to first create a **bucket**—also called a container—and then upload a binary **object** to that new bucket, as follows.

```
# Create new bucket
bucket = js.object_store.create('my_bucket_name')
# Create new object within bucket
buckobj = bucket[0].create_object('stuff')
fo = open('\path to your data\stuff.txt','rb')
# Upload file contents to new object
buckobj.upload(fo)
```

To verify that these actions worked, you can log into the OpenStack portal and check the current container state, as shown in figure 3.6 on the following page. The complete code is in notebook 6.

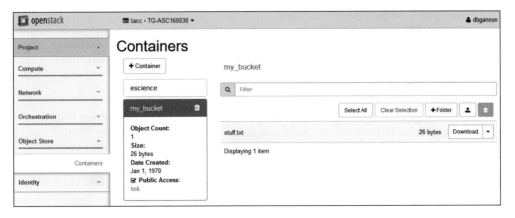

Figure 3.6: View of containers in the OpenStack object store.

3.6   Transferring and Sharing Data with Globus

When using cloud resources in science or engineering, we often need to copy data between cloud and non-cloud systems. For example, we may need to move genome sequence files from a sequencing center to the cloud for analysis, and analysis results to our laboratory. Figure 3.7 shows how Globus services can be used for such purposes. We see in this figure three different storage systems: one associated with a sequencing center, a cloud storage service, and one located on a personal computer. Each runs a lightweight Globus Connect agent that allows it to participate in Globus transfers.

> **The Globus Connect agent** enables a computer system to interact with the Globus file transfer and sharing service. With it, the user can easily create a Globus endpoint on practically any system, from a personal laptop to a cloud or national supercomputer. Globus Connect comes in two flavors: **Globus Connect Personal** for use by a single user on a personal machine, and **Globus Connect Server** for use on multiuser computing and storage resources.
>
> Globus Connect supports a wide variety of storage systems, including both various POSIX-compliant storage systems (Linux, Windows, MacOS; Lustre, GPFS, OrangeFS, etc.) and various specialized systems (HPSS, HDFS, S3, Ceph RadosGW via the S3 API, Spectra Logic BlackPearl, and Google Drive). It also interfaces to a variety of different user identity and authentication mechanisms.

Note that all of the public cloud examples presented earlier in this chapter involved client-server interactions: in each case, we had to run the data upload to the cloud from the machine that mounts the storage system with the data. In

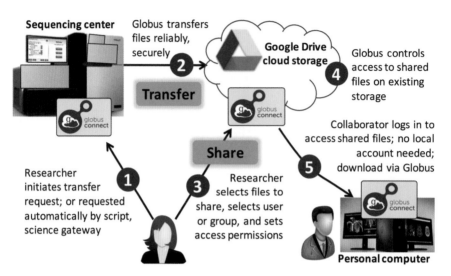

Figure 3.7: Globus transfer and sharing services used to exchange data among sequencing center, remote storage system (in this case, Google Drive), and a personal computer.

contrast, Globus allows third-party transfers, meaning that you can drive from a computer A a transfer from one endpoint B to another endpoint C. This capability is often important when automating scientific workflows.

The figure depicts a series of five data manipulation operations. (1) A researcher requests, for example via the Globus web interface, that a set of files be transferred from a sequencing center to another storage system, in this case the Google Drive cloud storage system. (2) The transfer then proceeds without further engagement by the requesting researcher, who can shut down her laptop, go to lunch, or do whatever else is needed. The Globus cloud service (not shown in the figure) is responsible for completing the transfer, retrying failed transfers if required, and notifying the user of failure only if repeated retries are not successful. The user requires only a web browser to access the service, can transfer data to and from any storage system that runs the Globus Connect software, and can authenticate using an institutional credential. Steps 3–5 are concerned with data sharing, which we discuss in section 3.6.2 on page 54.

3.6.1 Transferring Data with Globus

Figure 3.8 shows the Globus web interface being used to transfer a file. Cloud services are being used here in two ways: Data are being transferred from cloud storage, in this case Amazon S3; and the Globus service is cloud-hosted software as a service—running, as we discuss in chapter 14, on the Amazon cloud.

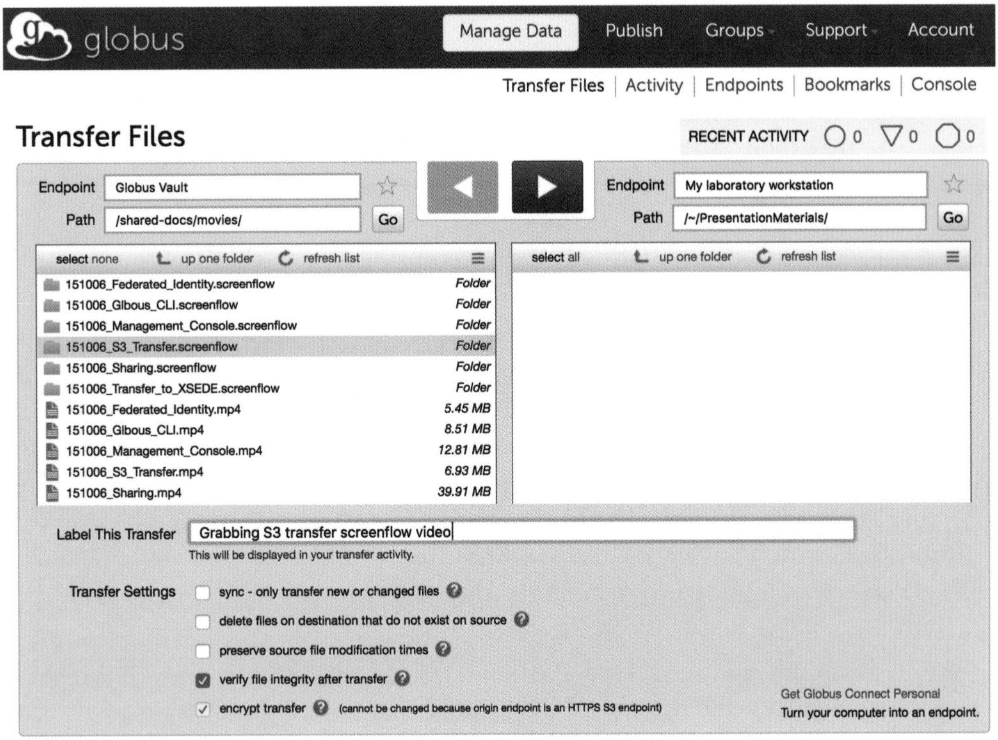

Figure 3.8: Globus transfer web interface. We have selected a single file on the Globus endpoint **Globus Vault** to transfer to the endpoint **My laboratory workstation**.

Globus also provides a REST API and a Python SDK for that API, allowing you to drive transfers entirely from Python programs. We use the code in figure 3.9 on page 55 to illustrate how the Python SDK can be used to perform and then monitor a transfer. The first lines of code, labeled (a), create a transfer client instance. This handles connection management, security, and other administrative issues. The code then (b) specifies the identifiers for the source and destination endpoints. Each Globus endpoint and user are named by a universally unique identifier (UUID). An endpoint's identifier can be determined via the Globus web client or programmatically, via an endpoint search API. For example, with the Python SDK you can write:

```
tc = globus_sdk.TransferClient(...)
for ep in tc.endpoint_search('String to search for'):
    print(ep['display_name'])
```

In figure 3.9, we hard-code the identifiers of two endpoints that the Globus team operates for use in tutorials. The code also specifies (c) the source and destination paths for the transfer.

Next, the code (d) ensures that the endpoints are activated. In order for the transfer service to perform operations on endpoint file systems, it must have a credential to authenticate to the endpoint as a specific local user. The process of providing such a credential to the service is called **endpoint activation** [22]. The `endpoint_autoactivate` function checks if the endpoint is activated for the calling user, or can be automatically activated using a cached credential that will not expire for at least a specified period of time. Otherwise it returns a failure condition, in which case the user can use the Globus web interface to authenticate and thus provide a credential that Globus can use for some time period. We show code that handles that case for the destination endpoint.

In (e)–(g), we assemble and submit the transfer request, providing the endpoint identifiers, the source and destination paths, and (since we want to transfer a directory) the recursive flag. In (h), we check for task status. This blocking call returns after a specified timeout or when the task terminates, whichever is sooner. In this case, we choose to (i) terminate the task if it has not completed by the timeout; we could instead repeat the wait.

More examples of the Globus Python SDK are in notebook 8. Globus also provides a command line interface (CLI), implemented via the Python SDK, that can be used to perform the operations just described. For example, the following command transfers a directory from one endpoint to another. More details on how to use the CLI are available online `docs.globus.org`.

```
globus transfer --recursive \
    "ddb59aef-6d04-11e5-ba46-22000b92c6ec":shared_dir   \
    "ddb59af0-6d04-11e5-ba46-22000b92c6ec":destination_directory
```

3.6.2   Sharing Data with Globus

Globus also makes it easy to share data with colleagues, as shown in figure 3.7 on page 52, steps 3–5. A **shared endpoint** is a dynamically created construct that enables a folder on an existing endpoint to be shared with others. To use this feature, you first create a shared endpoint, designating an existing endpoint and folder, and then grant read and/or write permissions on that shared endpoint to the Globus user(s) and/or group(s) that you want to be able to access it. Shared endpoints can be created and managed both via the Globus web interface (see figure 3.10 on page 56) and programmatically, as we show in chapter 11.

```python
# (a) Prepare transfer client
import globus_sdk
tc = globus_sdk.TransferClient()

# (b) Define the source and destination endpoints for the transfer
source_endpoint_id = 'ddb59aef-6d04-11e5-ba46-22000b92c6ec'
dest_endpoint_id = 'ddb59af0-6d04-11e5-ba46-22000b92c6ec'

# (c) Define the source and destination paths for the transfer
source_path = '/share/godata/'
dest_path = '/~/'

# (d) Ensure endpoints are activated
tc.endpoint_autoactivate(source_endpoint_id, if_expires_in=3600)
r = tc.endpoint_autoactivate(dest_endpoint_id, if_expires_in=3600)
while (r['code'] == 'AutoActivationFailed'):
    print('To activate endpoint, open URL in browser:')
    print('https://www.globus.org/app/endpoints/%s/activate'
          % dest_endpoint_id)
    # For python 2.X, use raw_input() instead
    input('Press ENTER after activating the endpoint:')
    r = tc.endpoint_autoactivate(ep_id, if_expires_in=3600)

# (e) Start transfer set up
tdata = globus_sdk.TransferData(tc, source_endpoint_id,
                                dest_endpoint_id,
                                label='My test transfer')

# (f) Specify a recursive transfer of directory contents
tdata.add_item(source_path, dest_path, recursive=True)

# (g) Submit transfer request
r = tc.submit_transfer(tdata)
print('Task ID:', submit_result['task_id'])

# (h) Wait for transfer to complete, with timeout
done = tc.task_wait(r['task_id'], timeout=1000)

# (i) Check for success; cancel if not completed by timeout
if done:
    print('Task completed')
else:
    cancel_task(r['task_id'])
    print('Task did not complete in time')
```

Figure 3.9: Using the Globus Python SDK to initiate and monitor a transfer request.

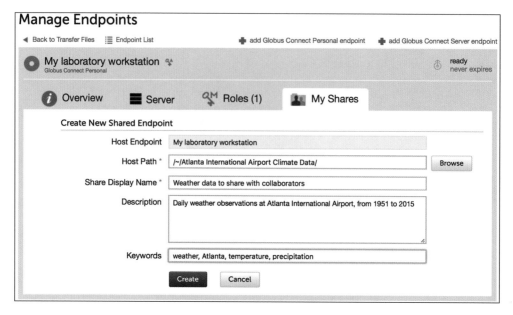

Figure 3.10: Globus web interface for creating a shared endpoint.

3.7 Summary

We have introduced in this chapter fundamental methods for interacting with cloud storage services. We focused, in particular, on blob services and NoSQL table services. As we noted in chapter 2, these are far from being the only cloud storage offerings. Indeed, as we show in the next chapter, POSIX file storage systems attached to virtual machines are particularly important for high-performance applications. So too are data analytic warehouses, which we discuss in part III.

We also delved more deeply in this chapter into cloud access methods first introduced in section 1.4 on page 8. We showed how cloud provider portals can be used for interactive access to resources and services, and also how Python SDKs can be used to script data management and analysis workflows. We used them, in particular, to script the task of uploading a set of data objects and building a searchable NoSQL table of metadata where each row of the table corresponds to one of the uploaded blobs of data.

As you studied the Python code that we provided for Amazon, Azure, and Google, you may have been frustrated that the programs for the different versions are almost but not quite the same. Why not have a single API and corresponding SDK for all clouds? As we have said, various attempts to create such a uniform

SDK are under way, but these attempts can cover only the common intersection of each cloud's capabilities. Each cloud service grew out of a different culture in a different company, and so are not identical. The resulting creative ferment has contributed to the explosion of tools and concepts that we describe in this book.

3.8 Resources

We provide all examples in this chapter as Jupyter notebooks, as described in chapter 17. You first need to install the SDKs for each cloud. The SDKs and links to documents are here:

- Amazon Boto3 `aws.amazon.com/sdk-for-python/`

- Azure `azure.microsoft.com/en-us/develop/python/`

- Google's Cloud `cloud.google.com/sdk/`

- Openstack CloudBridge `cloudbridge.readthedocs.io/en/latest/`

- The Globus Python SDK `github.com/globus/globus-sdk-python` and the Globus CLI `github.com/globus/globus-cli`

You also need an account on each cloud. Chapter 1 provides links to the portals where you can obtain a trial account.

Part II

Computing in the Cloud

Building your own cloud
What you need to know
Using Eucalyptus
Part IV Using OpenStack

Security and other topics
Securing services and data
Solutions
History, critiques, futures **Part V**

Part III **The cloud as platform**

Data analytics	Streaming data	Machine learning	Research data portals
Spark & Hadoop	Kafka, Spark, Beam	Scikit-Learn, CNTK,	DMZs and DTNs, Globus
Public cloud Tools	Kinesis, Azure Events	Tensorflow, AWS ML	Science gateways

Managing data in the cloud
File systems
Object stores
Databases (SQL)
NoSQL and graphs
Warehouses
Part I Globus file services

Computing in the cloud
Virtual machines
Containers – Docker
MapReduce – Yarn and Spark
HPC clusters in the cloud
Mesos, Swarm, Kubernetes
Part II HTCondor

Part II:
Computing in the Cloud

While scientists were first attracted to the cloud by the ability to store and share data, it was the introduction of cheap on-demand computing that created a paradigm shift. In this second part of the book, we follow the pattern established in the preceding one: we first introduce principles and then show how you can use both cloud portals and Python SDKs to compute on various cloud platforms.

Computing in the cloud has gone through a fascinating evolution. It started with virtualization, an old computing technology first invented in the context of mainframe computers and later adopted within data centers as a means of allowing customers to create environments and services that are uniquely tailored to their needs. Virtual machines can be started and stopped easily, and the customer is charged only for the time that the machine instance is running. In chapter 5, we describe how to create and manage virtual machines on cloud platforms.

A second stage of the evolution of computing in the cloud was the introduction of containers as a means of encapsulating software. Container technologies allow researchers to share deployed applications that can be deployed rapidly on any cloud and then run with a single command. In chapter 6, we show you how to create and deploy containers based on a technology called Docker.

Scale has always been a critical cloud capability and a major requirement of scientists. By "scale" we mean the ability of computation to be spread over multiple cloud servers to exploit parallelism in the application. In chapter 7, we consider four types of parallel application execution:

- **SPMD clusters** in the cloud, for traditional HPC-style computing.

- **Many task** or **high throughput** parallel computation, characterized by a large bag of tasks with few or no dependencies and that thus can be executed in parallel.

- **MapReduce and BSP** style parallelism, in which a single thread of control applies parallel operators over distributed data. In the cloud, such computations often involve executing a directed graph of data parallel operations. This model is used in tools such as Spark, many of the open source data analytics tools, and most of the deep learning systems that we discuss in part III.

- **Microservices**, the most "cloud native" computational model. It uses frameworks such as Mesos and Kubernetes to allow applications to be composed of swarms of dockerized, mostly stateless, small communicating services.

We also include a short discussion of **serverless computing**, a relatively new capability of the big public clouds that, like many computational ideas, has deep roots in operating system design. Briefly stated, it allows a programmer to define the code for an application plus the events that should cause that code to execute, and then release the code to the cloud in such a way that its execution does not require any cloud resource deployment by the user or programmer.

Chapter 4

Computing as a Service

"In pioneer days they used oxen for heavy pulling, and when one ox couldn't budge a log, they didn't try to grow a larger ox. We shouldn't be trying for bigger computers, but for more systems of computers."

—Admiral Grace Hopper

The two simplest forms of cloud computing are perhaps the best known to scientists and engineers: storage and computing on demand. We covered storage in part I of the book; we turn next to cloud computing: computing as a service.

A cloud can provide near-instant access to as much computing and storage as you need, when you need it, along with the ability to obtain computers with the specific configuration(s) that you want. Furthermore, these capabilities are all just an API call or a few mouse clicks away. As we will see, computing services can be delivered in several different ways. The most basic is known in the cloud industry as **infrastructure as a service** (IaaS) because it provides virtualized infrastructure to its users. To understand what IaaS compute looks like, let's consider some typical scenarios.

- You need 100 CPUs now, rather than when your job reaches the head of the queue in your lab cluster.

- You need to access a GPU cluster or a computer with 1 TB memory, but such a resource does not exist in the lab.

- You need computers loaded with 10 different Linux variants to test the portability of your new software prior to distribution.

- You want a machine outside your firewall that you can share with your collaborators for a project.

Each scenario can be addressed in multiple ways by cloud computing. The question we look at in this chapter is how to determine the best approach and how to evaluate the pros and cons of picking a solution.

4.1 Virtual Machines and Containers

Public cloud data centers comprise many thousands of individual servers. Some servers are used exclusively for data services and supporting infrastructure and others for hosting your computations. When you compute in the cloud, you do not run directly on one of these servers in the way that you would in a conventional computational cluster. Instead, you are provided with a virtual machine running your favorite operating system. A **virtual machine** is just the software image of a complete machine that can be loaded onto the server and run like any other program. The server in the data center runs a piece of software called a **hypervisor** that allocates and manages the server's resources that are granted to its "guest" virtual machines. In the next chapter, we delve into how virtualization works, but the key idea is that when you run in a VM, it looks exactly like a server running whatever operating system the VM is configured to run.

For the cloud operator, virtualization has huge advantages. First, the cloud provider can provide dozens of different operating systems packaged as VMs for the user to choose from. To the hypervisor, all VMs look the same and can be managed in a uniform way. The cloud management system (sometimes called the **fabric controller**) can select which server to use to run the requested VM instances, and it can monitor the health of each VM. If needed, the cloud monitor can run many VMs simultaneously on a single server. If a VM instance crashes, it does not crash the server. The cloud monitor can record the event and restart the VM. User applications running in different VMs on the same server are largely unaware of each other. (A user may notice another VM when they impact the performance or response of their VM.)

We provide in chapter 5 detailed instructions on how to deploy VMs on the Amazon and Azure public clouds, and on OpenStack private clouds.

Containers are similar to VMs but are based on a different technology and serve a slightly different purpose. Rather than run a full OS, a container is layered on top of the host OS and uses that OS's resources in a clever way. Containers allow you to package up an application and all of its library dependencies and data into a single, easy-to-manage unit. When you launch the container, the application can be configured to start up, go through its initialization, and be running in seconds. For example, you can run a web server in one container and a database server in another; these two containers can discover each other and communicate as needed. Or, if you have a special simulation program in a container, you can start multiple instances of the container on the same host.

Containers have the advantage of being extremely lightweight. Once you have downloaded a container to a host, you can start it and the application(s) that it contains quasi-instantly. Part of the reason for this speed is that a container instance can share libraries with other container instances. VMs, because they are complete OS instances, can take a few minutes to start up. You can run many more containers on a single host machine than you can effectively run the same number of VMs. Figure 4.1 illustrates the difference between the software stack of a server running multiple VMs versus a server running a single OS and multiple containers on a typical server in a cloud.

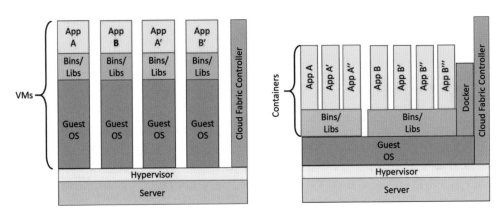

Figure 4.1: Virtual machines vs. containers on a typical cloud server.

Building a container to run a single application is simple compared with the task of customizing a VM to run a single application. All you need to do is create a script that identifies the needed libraries, source files, and data. You can then run the script on your laptop to test the container, before uploading the container to a repository, from where it can be downloaded to any cloud. Importantly, containers are completely portable across different clouds. In general, VM images cannot be ported from one cloud framework to another.

Containers also have downsides. The most serious issue is security. Because containers share the same host OS instance, two containers running on the same host are less isolated than two VMs running on that host. Managing the network

Table 4.1: Virtual machines and containers, compared.

Virtual machines	Containers
Heavyweight	Lightweight
Fully isolated; hence more secure	Process-level isolation; hence less secure
No automation for configuration	Script-driven configuration
Slow deployment	Rapid deployment
Easy port and IP address mapping	More abstract port and IP mappings
Custom images not portable across clouds	Completely portable

ports and IP addresses used by containers can be slightly more confusing than when working with VMs. Furthermore, containers are often run on top of VMs, which can exacerbate the confusion.

In chapter 6, we work though examples of using containers in some detail, describing in particular the Docker container system `docker.com`, how to run containers, and how to create your own containers.

Table 4.1 lists some of the the pros and cons of the virtual machine and container approaches. Needless to say, both technologies are evolving rapidly.

4.2 Advanced Computing Services

Cloud vendors such as Amazon, Microsoft, and Google have many additional services to help your research, including special data analysis clusters, tools to handle massive streams of events from instruments, and special machine learning tools. We discuss these services in part III of the book.

A common issue of concern to scientists and engineers is **scale**. VMs and containers are a great way to virtualize a single machine image. However, many scientific applications require multiple machines to process many data or to perform a complex simulation. You may already know how to run parallel programs on clusters of machines and you now want to know whether you can run those same programs on the cloud. The answer depends on the specifics of the application. Most high-performance parallel applications are based on the **Message Passing Interface** (MPI) standard [147]. As we describe in chapter 7, Amazon and Azure provide an extensive set of tools for building Linux MPI clusters.

Cloud users can also exploit parallelism in other ways. For example, **many task** (MT) parallelism [221] is used to tackle problems in which you have hundreds of similar tasks to run, each(largely) independent of the others. Another method is called **MapReduce** [108], made popular by the Hadoop computational framework

[258]. MapReduce is related to a style of parallel computing known as **bulk synchronous parallelism** (BSP) [139]. We cover these topics, also, in chapter 7.

A compelling feature of the cloud is that it provides many ways to create highly scaled applications that are also interactive. The **Spark** system [265], originally developed at University of California Berkeley, is more flexible than Hadoop and is a form of BSP computing that can be used interactively from Jupyter. Google has released a service called **Cloud Datalab**, based on Jupyter, for interactive control of its data analytics cloud. The Microsoft Cloud Business Intelligence (**Cloud BI**) tool supports interactive access to data queries and visualization. We discuss these tools in chapter 8.

Managing your cloud computing resources can become complicated when you need to scale beyond a few VMs or containers. Keeping track of many processes spread over many cloud VMs is not easy. Fortunately, several new tools have been adopted by the public clouds to help with this challenge. For managing large numbers of containers, you can use the Docker **Swarm** tools `docker.com/products/docker-swarm` and Google's **Kubernetes container management** [26, 79] (which Google uses for its own container management). Many people use the venerable **HTCondor** system [243] to manage many task parallel computation. (HTCondor is used in the Globus Genomics system that we describe in chapter 11.) **Mesos** [154] provides another distributed operating system with a web interface that allows you to manage many applications in the cloud at the same time. All of these systems are already available, or can easily be deployed, on cloud platforms. We describe them in chapter 7.

One other computation service programming model that is common in cloud computing is **dataflow**. This model plays a significant role in the analysis of streaming data, as discussed in chapter 9.

4.3 Serverless Computing

An interesting recent trend in cloud computing is the introduction of **serverless computing** as a new paradigm for service delivery. As we show in the chapters ahead, computation and data analysis can be deployed in the cloud via a range of special services. In the majority of cases, the user must deploy VMs, either directly or indirectly, to support these capabilities. Doing so takes time, and the user is responsible for deleting the VMs when they are no longer needed. At times. however, this overhead is not acceptable, such as when you want an action to take place in response to a relatively rare event. The cost of keeping a VM running continuously so that a program can wait for the event may be unacceptably high.

Serverless computing is similar to the old Unix concepts of a daemon and cron jobs, whereby a program is managed by the operating system and is executed only when specific conditions arise. In serverless computing, the user provides a simple function to be executed, again under certain conditions. For example, the user may wish to perform some bookkeeping when a new file is created in a cloud repository or to receive a notification when an important event occurs. The cloud provider keeps a set of machines running to execute these functions on the user's behalf; the user is charged only for the execution of the task, not for maintaining the servers. We return to this topic in chapters 9 and 18.

4.4 Pros and Cons of Public Cloud Computing

Public cloud computing has both pros and cons. Important advantages include the following.

- *Cost*: If you need a resource for only a few hours or days, the cloud is much cheaper than buying a new machine.

- *Scalability*: You are starting a new lab and want to start with a small number of servers, but as your research group grows, you want to be able to expand easily without the hassle of managing your own racks of servers.

- *Access*: A researcher in a small university or an engineer in a small company may not even have a computer room or a lab with room for racks of machines. The cloud may be the only choice.

- *Configurability*: For many scientific disciplines, you can get complete virtual machines or containers with all the standard software you need pre-installed.

- *Variety*: Public cloud systems provide access to a growing diversity of computer systems. Amazon and Azure each provide dozens of machine configurations, ranging from a single core with a gigabyte of memory to multicore systems with GPU accelerators and massive amounts of memory.

- *Security*: Commercial cloud providers have excellent security. They also make it easy to create a virtual network that integrates cloud resources into your network.

- *Upgradeability*: Cloud hardware is constantly upgraded. Hardware that you buy is out of date the day that it is delivered, and becomes obsolete quickly.

- *Simplicity*: You can manage your cloud resources from a web portal that is easy to navigate. Managing your own private cluster may require sophisticated system administration skills.

Disadvantages of computing as a service include the following.

- *Cost*: You pay for public cloud by the hour and the byte. Computing the total cost of ownership of a cluster of machines housed in a university lab or data center is not easy. In many environments, power and system administration are subsidized by the institution. If you need to pay only for hardware, then running your own cluster may be cheaper than renting the same services in the cloud. Another accounting oddity, perhaps peculiar to the U.S., is that some universities charge overhead on funds obtained from a federal funding source for public cloud but not when equivalent funds are used for hardware purchases.

  Academic researchers may also have the option of accessing a national facility such as Jetstream, Chameleon, or the European science cloud. The cost here is the work involved in writing a proposal.

- *Variety*: The cloud does not provide every type of computing that you may require, at least not today. In particular, it is not a proper substitute for a large supercomputer. As we show in chapter 7, both Amazon and Azure support the allocation of fairly sophisticated high performance computing (HPC) clusters. However, these clusters are not at the scale or performance level of the top 500 supercomputers.

- *Security*: Your research concerns highly sensitive data, such as medical information, that cannot be moved outside your firewall. As stated above, there are ways to extend your network into the cloud, and the cloud providers provide HIPAA-compliant facilities. However, the paperwork involved in getting approval to use these solutions may be daunting.

- *Dependence* on one cloud vendor (often referred to as vendor lock-in). This situation is changing, however. As the public clouds converge in many of their standard offerings and compete on price, moving applications between cloud vendors has become easier.

Another way to think about the cost of computing is to weigh the cost of computing for an *individual*, who may use only a few hours of computing per week, versus for an entire *institution*, such as a university or large research center, which may in aggregate use many tens of thousands of hours per week. One sensible solution is for the institution to sign a long-term contract with a public cloud provider that allows the institution to pick up the tab for its researchers. There are several ways to make this approach economically attractive to both the institution and the cloud provider. For example, universities can negotiate cloud access as part of a package deal involving software licenses or institutional data back-up.

An institution may also have the resources and expertise required to build a private mini-data center, in which it can deploy an OpenStack cloud that it makes available to all of its employees. Depending on the institution and its workloads, this approach may be more cost effective than others. It may also be required for data protection reasons. This approach leaves open the possibility of a hybrid solution, in which you spill over to the public cloud when the private cloud is saturated. This is often referred to as **cloud bursting**. This hybrid solution has become a common model for many large corporations and is well supported by cloud providers such as Microsoft and IBM. Aristotle `federatedcloud.org` is an example of an academic cloud that supports cloud bursting.

4.5   Summary

Clouds can provide computing resources as a service, at scales ranging from a single virtual machine with one virtual core and a few gigabytes of memory to full HPC clusters. The type and scale of service you use depend on the nature of the application. The following are examples.

- You may need only an extra Linux or Windows server to run an application, and you do not want to saddle your laptop with the extra load. In this case, a single VM running on a multicore server with a large memory is all you may need. You deploy it in a few minutes and remove it when you are done.

- You want to run a standard application such as a database to share with other users. In this case, the easiest solution is to run a containerized instance of your favorite database on a VM designed to run containers or on a dedicated container hosting service.

- You have an MPI-based parallel program that does not need thousands of processors to get the job done in a reasonable amount of time. In this case the public clouds have simple tools to spin up a HPC cluster that you can use for your job.

- You have a thousand tasks to run that produce data you need to analyze interactively. This may be a job for Spark with a Jupyter front end or Hadoop if it can be rendered as a MapReduce computation.

- If the thousand-task computations are more loosely coupled and can be widely distributed, then HTCondor is a natural choice.

- If you are processing large streams of data from external sources, a dataflow stream processing tool may be the best solution.

Other considerations, notably cost and security, can also enter into your choice of resource and system. We discuss security in chapter 15.

4.6 Resources

Each of the major public cloud vendors provides excellent tutorials for using its computing services. In addition, entire books have been written on each topic covered in this chapter. Four that we particularly like are: *Programming AWS EC2* by Jurg van Vliet and Flavia Paganelli [251]; *Amazon Web Services in Action*, by Andreas Wittig and Michael Wittig [263]; *Programming Google App Engine with Python*, by Dan Sanderson [231]; and *Microservices, IoT and Azure: Leveraging DevOps and Microservice Architecture to deliver SaaS Solutions* by Bob Familiar [119]. We point to additional resources in the chapters that follow.

Chapter 5

Using and Managing Virtual Machines

"We all live every day in virtual environments, defined by our ideas."
                                                        —Michael Crichton

The introduction by Amazon of its Elastic Compute Cloud (EC2) service in 2006 marked the true beginning of cloud computing. EC2 is based on virtualization technology, which allows one server to run independent, isolated operating systems (OSs) for multiple users simultaneously. Since then Microsoft, Google, and many others have introduced virtual machine (VM) services based on this technology.

In this chapter, we first provide a brief introduction to virtualization technology, and then proceed to describe how to create and manage VMs in the cloud. We start with creating a VM on EC2 and show how to attach an external disk. We then describe Microsoft's solution, Azure, and show how to create VM instances via both the Azure portal and a Python API.

The open source community has also been active in this space. Around 2008 three projects—Eucalyptus from the University of California Santa Barbara [212], OpenNebula from the Complutense University of Madrid [202], and Nimbus from the University of Chicago [191]—released cloud software stacks. Later NASA, in collaboration with Rackspace, released OpenStack, which is widely supported. We describe in this chapter one OpenStack-based system called Jetstream, a facility funded by the U.S. National Science Foundation, and show how to create OpenStack VMs on Jetstream. We provide additional information on Eucalyptus and OpenStack in chapters 12 and 13, respectively.

5.1 Historical Roots

Any modern computer has a set of basic resources: CPU data registers, memory addressing mechanisms, and I/O and network interfaces. The programs that control the computer are just sequences of binary codes corresponding to instructions that manipulate these resources, for example to `ADD` the contents of one register to the contents of another.

There are also important instructions for performing context switches, in which the computer stops executing one program and starts executing another. These state management instructions plus the I/O instructions are termed *privileged*. Such instructions are usually directly executed only by the OS, because you do not want users to be able to access state associated with other computations.

The OS has the ability to allow user programs (encapsulated as processes) to run the unprivileged instructions. But as soon as the user program attempts to access an I/O operation or other privileged instruction, the OS traps the instruction, inspects the request, and, if the request proves to be acceptable, runs a program that executes a safe version of the operation. This process of providing a version of the instruction that looks real but is actually handled in software is called **virtualization**. Other types of virtualization, such as virtual memory, are handled directly by the hardware with guidance from the OS.

In the late 1960s and early 1970s, IBM and others created many variations on virtualization and eventually demonstrated that they could virtualize an entire computer [104]. What resulted was the concept of a **hypervisor**: a program that manages the virtualization of the hardware on behalf of multiple distinct OSs. Each such OS instance runs on its own complete VM that the hypervisor ensures is completely isolated from all other instances running on the same computer. Here is an easy way to think about it. The OS allows multiple user processes to run simultaneously by sharing the resources among them; the hypervisor below the OS allows multiple OSs to share the real physical hardware and run concurrently. Many hypervisors are available today, such as Citrix Xen, Microsoft Hyper-V, and VMWare ESXi. We refer to the guest OSs running on the hypervisors as VMs. Some hypervisors run on top of the host machine OS as a process, such as VirtualBox and KVM, but for our purposes the distinction is minor.

While this technical background is good for the soul, it is not essential to learning how to create and manage VMs in the cloud. In the remainder of this chapter we dig into the mechanics of getting science done with VMs. We assume that you are familiar with Linux and focus in our examples on creating Linux VMs. This choice does not imply that Windows is not available. In fact, all three public clouds we talk about in this book allow you to create VMs running Windows just as easily as Linux. So if you need a Windows VM for some of your work, rest assured that almost everything that we present works for Windows, too. We try to point out the occasional exceptions to this rule.

Each of the three public clouds and the NSF Jetstream cloud has a web portal that guides you through the steps needed to create and manage VM instances. If you have never used a cloud, you are well advised to start there. We introduce selected interfaces and describe how to get started with each.

5.2   Amazon's Elastic Compute Cloud

We describe first how to create VM instances on Amazon's Elastic Compute Cloud service and then how to attach storage to our VMs.

5.2.1   Creating VM Instances

We start on the Amazon portal at `aws.amazon.com`, where we can log in or create an account. Figure 5.1 shows what we see when we log in. We are interested in VMs, so we click on `EC2` or `Launch a Virtual Machine`. This brings us to another series of views, with instructions on how to launch a basic "Amazon Linux" instance. We can then specify our desired host service `Instance Type`, which determines the number of cores that our VM is to use, the required memory size, and network performance. Literally dozens of choices exist, ordered from small to large, and priced accordingly (more on this below).

One important step during the launch process involves providing a key pair: the cryptographic keys that you use to access your running instance. If this is your first experience with EC2, you may be asked to create a key pair early in the process. You should do so. Give it a name and remember it. You then download the *private key* file to a secure place on your laptop where you can access it again. The corresponding *public key* is stored with Amazon. Just before you launch your instance, it asks you which key pair you want to use. After you select it, the public key is loaded into the instance. The other important choices involve storage options and security groups. We return to those later. Once you launch the instance you can monitor its status, as shown in figure 5.2 on the next page, where you see two stopped instances and one newly launched instance. The *Status Checks* shows that the new instance is still initializing. After a few moments, its status changes to a green check mark to indicate that the instance is ready to launch.

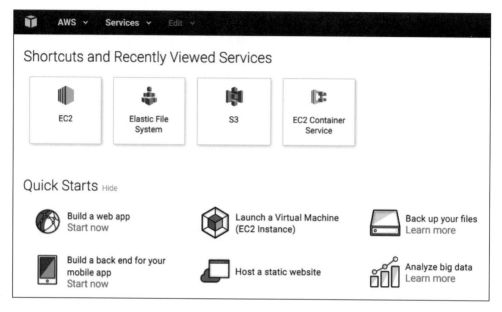

Figure 5.1: First view of the Amazon portal.

Figure 5.2: Portal instance view.

To connect to your instance, you need to use a secure shell command. On Windows the tool to use is called *PuTTY*. You need a companion tool called *PuTTYgen* to convert the downloaded private key into one that can be consumed by PuTTY. When you launch it, you use `ec2-user@IPAddress`, where the *IPAddress* is the IP address you can find in the Portal Instance View. The PuTTY *SSH* tab has an *Auth* tab that allows you to upload your converted private key. On a Mac or Linux machine, you can go directly to the shell and execute:

```
ssh -i path-to-your-private-key.pem ec2-user@ipaddress-of-instance
```

The following listing uses the Python Boto3 SDK to create an Amazon EC2 VM instance. It creates an `ec2` resource, which requires your `aws_access_key_id` and `aws_secret_access_key`, unless you have these stored in your `.aws` directory. It then uses the `create_resources` function to request creation of the instance.

```
import boto3
ec2 = boto3.resource('ec2', 'us-west-2')
ec2.create_instances(ImageId='ami-7172b611', 't2.micro',
                    MinCount=1, MaxCount=1)
```

The `ImageId` argument specifies the VM image that is to be started and the `MinCount` and `MaxCount` arguments the number of instances needed. (In this case, we want five instances, but we will accept a single instance if that is all that is available.) Other optional arguments can be used to specify instances. For example, the *instance type*: Do you want a small virtual computer, with limited memory and computing power, or a big one with many cores and lots of storage? (As we discuss below, you pay more for the latter.) Having created the instance(s), we define and call a `show_instances` function that uses `instances.filter` to obtain and display a list of running instances. The last line shows the result.

```
# A function that lists instances with a specified status
def show_instance(status):
    instances = ec2.instances.filter(
        Filters=[{'Name':'instance-state-name','Values':[status]}])
    for instance in instances:
        print(instance.id, instance.instance_type,
            instance.image_id, instance.public_ip_address)

show_instance('running')
('i-0a184b56b0ebdba98', 't2.micro', 'ami-7172b611', '146.137.70.71')
```

Notebook 7 provides more examples, showing, for example, how to suspend and terminate instances, check instance status, and attach a virtual disk volume.

5.2.2 Attaching Storage

We now discuss the three types of storage that, as noted in chapter 2, can be attached to a VM: instance storage, Elastic Block Store, and Elastic File System. **Instance storage** is what comes with each VM instance. It is easy to access, but when you destroy a VM, all data saved in its instance storage goes away.

We allocate **Elastic Block Store (EBS)** storage independent of a VM and then attach it to a running VM. EBS volumes persist and thus are good for databases and other data collections that we want to keep beyond the life of a

VM. Furthermore, they can be encrypted and thus used to hold sensitive data. To create an EBS volume, go to the *volumes* tab of the EC2 Management console and click ⎡Create Volume⎤. The dialog box in figure 5.3 allows you to specify volume size (20 GB here), encryption state, *snapshot ID*, and *availability zone*.

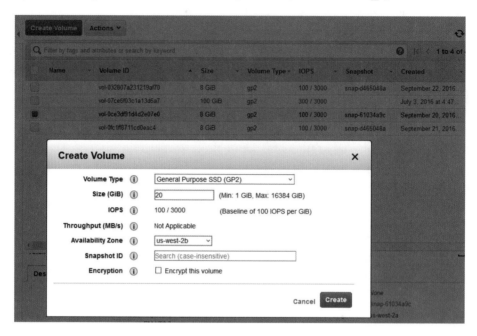

Figure 5.3: Creating an EBS volume.

We selected the `us-west-2b` availability zone because we want to attach the volume to the instance created earlier. The easiest way to make the attachment is through the *Actions* tab in the Volume management console. However, you can do much of the volume attaching and mounting in Python. First let us look at the list of our current volumes. The following is a transcript of an IPython session.

```
In [3]: vols = ec2.volumes.filter(Filters=[])
In [4]: for vol in vols:
            print(vol.id, vol.size, vol.state)
        ('vol-032807a231219af70', 8, 'in-use')
        ('vol-0bdd0584d0833e691', 20, 'available')
        ('vol-07ce6f03c1a13d5a7', 100, 'in-use')
        ('vol-0ce3df91d4d2e07e0', 8, 'in-use')
        ('vol-0fc1ff8711cd0eac4', 8, 'in-use')
```

We see that the 20 GB volume that we created in the portal session above is *available*, so let's attach it to our instance, as follows.

```
In [5]: vol = ec2.Volume('vol-0bdd0584d0833e691')
In [6]: vol.attach_to_instance(InstanceId='i-0a184b56b0ebdba98',
                               Device='/dev/xvdh' )
        {u'AttachTime': datetime.datetime(2016,9,23,18,15,49,308000,
                                           tzinfo=tzutc()),
         u'Device': '/dev/xvdh',
         ... more attach metadata not shown ...
        }
```

We must next mount the volume. We cannot use Boto3 for that task, but must log into the instance and issue commands to the instance OS. If we have many instances, we may want to script this task in Python. We can use the `ssh` command to feed the mount commands to the instance directly. We first define a helper function to invoke `ssh` and pass a script, as follows.

```
In [7]: def myexec( pathtopem, hostip, commands):
    ssh = subprocess.Popen(['ssh', '-i', pathtopem,
               'ec2-user@%s'%hostip, commands ],
        shell=False, stdout=subprocess.PIPE, stderr=subprocess.PIPE)
    result = ssh.stdout.readlines()
    if result == ['error']:
        error = ssh.stderr.readlines()
        print >>sys.stderr, "ERROR: %s" % error
        return "error"
    else:
        return result
```

This function requires the path to your private key, the IP address of the instance, and the command script as a string. To do the mount, we can call this function as follows. We first create a file system on our new virtual device. Next we make a directory `data` at the root level and then mount the device at that point. To verify that the job got done, we do a `df` to display the free space.

```
In [8]: command = 'sudo mkfs -t ext3 /dev/xvdh\n \
               sudo mkdir /data\n \
               sudo mount /dev/xvdh /data\n \
               df\n '
In [9]: myexec('path-to-pem-file', 'ipaddress', command)
    [output from the mkfs command ... ]
    'Filesystem     1K-blocks     Used Available Use%Mounted on\n',
    '/dev/xvda1      8123812 1211120   6812444  16% /\n',
    'devtmpfs         501092       60    501032   1%/dev\n',
    'tmpfs            509668        0    509668   0%/dev/shm\n',
    '/dev/xvdh       20511356    45124  19417656   1% /data\n'
```

We have now successfully created and mounted our 20 GB EBS volume on the new instance. One shortcoming of EBS storage is that it can be mounted only on one instance at a time. However, we can detach an EBS volume from one instance and then reattach it to a different instance in the same availability zone.

If you want a volume to be shared with multiple instances, then you can use the third type of instance storage, called Elastic File System, that implements the Network File System (NFS) standard. This takes a few extra steps to create and mount, as shown in the companion Jupyter notebook.

How much do you want to pay? When it comes to paying for cloud computing, public cloud services present a bewildering range of options. On Amazon, for example, these range from less than a cent per hour for a *nano* instance to several dollars per hour for a big-storage or graphical processing unit (GPU) system. And that is just for **on-demand instances**, instances that you request when you need them and pay for by the hour. **Reserved instances** provide lower costs (by up to 75%) when you reserve for between one and three years. And **spot instances** allow you to bid on spare Amazon EC2 computing capacity. You indicate the price that you are prepared to pay, and if Amazon has unused instances and your bid is above the current bid price, you get the machines that you asked for—with the proviso that if your bid price is exceeded during the lifetime of your instances, the instances are terminated and any work executing is lost. Spot prices vary considerably; but you can save a lot of money in this way, especially if your computations are not urgent.

Further complicating things is the fact that prices for different instance types, and especially for spot instances, can vary across Amazon regions. Thus, a really cost-conscious cloud user might be tempted to search across different instance types and regions for the best deal for a particular application. That would be a time-consuming process if you had to do it yourself, but researchers have built tools to do just that. Ryan Chard, for example, has developed a cost-aware elastic provisioner that can reduce costs by up to 95% relative to a less sophisticated approach [90].

5.3 Azure VMs

Microsoft's VM service was announced as Windows Azure in 2008, released to the public in 2010, extended to support Linux and its Python API in 2012, and rebranded as Microsoft Azure in 2014. As in the other public clouds, launching and managing a VM on Azure via the portal is straightforward. As seen in figure 5.4 on the following page, you have many VMs to choose from. Of special interest to us is the "Linux Data Science VM," which contains the R server, Anaconda Python, Jupyter, Postgres database, SQL server, Power BI desktop, and Azure command line tools, and many machine learning tools.

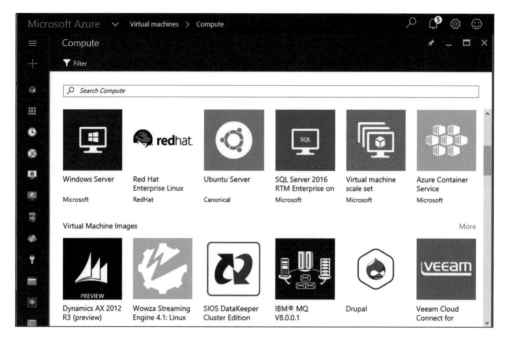

Figure 5.4: Azure portal VM launch page.

As with Amazon, one can write a Python script to launch this VM using the Azure Python SDK introduced in chapter 3. However, another SDK for Azure Python VM management, called **Simple Azure** [179] by Hyungro Lee, allows you to launch the Data Science VM as follows. (Once the status check returns Succeeded, your image is running, and you can connect to it.)

```
import simpleazure.simpleazure as sa
a = sa.SimpleAzure()
a.get_config()  # loads your credentials
img = a.get_registered_image(name="Azure-Data-Science-Core")
a.set_image(image = img)
a.set_location("West Europe")
a.create_vm()

#now check status
vars(a.get_status())
```

Using Simple Azure, one can also easily launch an *IPython* cluster and control it from Jupyter. Details are provided in the Simple Azure documentation [179].

81

5.4 Google Cloud VM Services

You can create VM instances from Google's Cloud portal in a manner similar to Azure and Amazon. A command line interface can also be used to manage VMs, but the Google Cloud Python SDK does not provide a way to launch a VM from a Python script. Google instead provides a unique system for running Python applications in the cloud called **AppEngine**. AppEngine applications are typically web services that scale automatically, a capability that can be useful for some scientific applications. We defer discussion of Google's compute offerings to chapter 7, where we discuss Google's Kubernetes and Cloud Container Service.

5.5 Jetstream VM Services

The Jetstream cloud uses a web interface called Atmosphere that allows users to select and run VM images in a manner similar to that supported by the public clouds. The primary difference is that the list of images available includes many that are packaged specifically for the science community. Here are a few examples.

- CentOS 6 with the MATLAB system and tools preinstalled with licenses available from Jetstream

- Bio-Linux 8, which adds more than 250 bioinformatics packages to an Ubuntu Linux 14.04 LTS base, providing around 50 graphical applications and several hundred command line tools

- The Accurate Species TRee ALgorithm (ASTRAL) phylogenetics package

- CentOS RStudio, which includes Microsoft R Open and MKL (Rblas)

- Wrangler iRODS 4.1 and a setup script for easy generation of the iRODS client environment on XSEDE resources

- Docker, the platform for launching Docker containers

- EPIC Modeling and Simulations: Explicit Planetary Isentropic-Coordinate (EPIC) Atmospheric Model Based on Ubuntu 14.04.3

Numerous Ubuntu and CentOS distributions are also available with various software development tools.

Galaxy, the gold standard for bioinformatics toolkits, is available and widely used on Jetstream. The Galaxy server comes preconfigured with hundreds of tools and commonly used reference datasets. The Galaxy team at the Johns Hopkins University, which hosts the main Galaxy server, is able to offload user jobs to instances running on Jetstream. Approximately 200 Galaxy VM instances are running on Jetstream at any one time. The Galaxy project wiki [20] describes how to deploy the latest Galaxy image.

To create a VM on Jetstream, you must first request an account and allocation via the XSEDE allocation process `www.xsede.org/allocations`. Once that is assigned, you can sign on to the system's web portal where you see a page describing your allocation. The banner at the top is illustrated in figure 5.5.

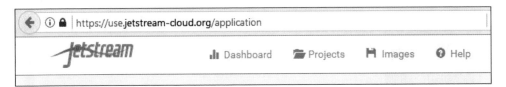

Figure 5.5: Banner for the Jetstream Atmosphere top level page.

You should first select the Projects tab in the banner, which allows you to access your current project or create a new one. Once you have a project, you can select it and you see a list of the instances, data volumes, and images that you have created with that project. By selecting the "NEW" button you can search for the image that is most interesting to you.

Figure 5.6 shows the Jetstream search function being used to locate an image with Docker installed. Once you select an image and click the (Launch Instance) button, you are prompted for the machine configuration. You can then monitor your deployment on your project page, as shown in figure 5.7, where we see two suspended small instances and a new medium-size instance.

5.6   Summary

We have described how to configure and launch a single VM on our candidate clouds. Each system provides a portal that makes configuration easy. The public clouds also provide command line SDKs that can be used to create and manage VMs programmatically. Python SDKs also exist for Amazon, Azure, and OpenStack. Google has an SDK for its AppEngine, which we do not cover here. We have found the Amazon Python SDK for launching VMs and managing the storage to be comprehensive. So, too, is the Azure SDK, but the Simple Azure package (which builds on the standard Azure SDK) makes launching VMs on Azure particularly

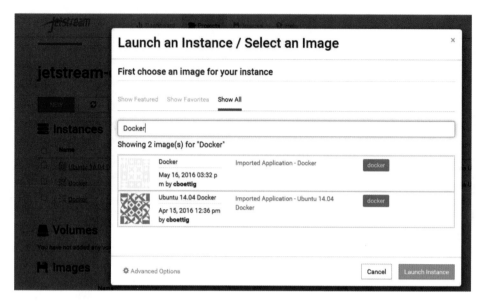

Figure 5.6: Search and launch functions for a VM image on Jetstream.

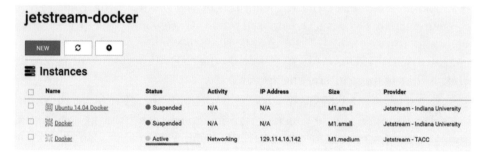

Figure 5.7: Project page showing the progress of the VM deployment.

simple. Jetstream's Atmosphere web interface makes launching VMs easy. The CloudBridge Python SDK for OpenStack can also be used: see notebook 6.

5.7 Resources

All the code samples described here are packaged as Jupyter notebooks, as described in chapter 17.

Chapter 6

Using and Managing Containers

"To see a world in a grain of sand ..."
—William Blake, *Auguries of Innocence*

Containers have become one of the most interesting and versatile alternatives to virtual machines for encapsulating applications for cloud execution. They enable you to run an application without source code modification or even recompilation on your Windows PC, your Mac, or any cloud. In this chapter we look at the history of their development, the key ideas that make them work, and of course how to use them on clouds. We focus on **Docker** container technology, because it is the most widely known and used, is easy to download and install, and is free.

We begin with some basics about containers. We show how to install Jupyter with Docker, and we mention other scientific applications that have been containerized. We then provide a simple example of how to create your own container that you can then run in the cloud or on your laptop.

6.1 Container Basics

People used to think that the best (indeed, in many cases, only) way to encapsulate software for deployment in the cloud was to create a VM image. That image could be shared with others, for example by placing it in a repository for Amazon images or in the Microsoft VM depot. Anybody could then deploy and run the image within an appropriate data center. However, not all virtualization tools are the same, so running a VM from Amazon on Azure or other clouds was a real problem. Endless debates arose about the merits and demerits of this situation, which usually went something like this: "This is just another form of evil vendor lock-in!" A great deal of thought was given to finding ways to address this evilness.

Meanwhile, others realized that the Linux kernel had some nice features that could be used to bound and contain the resource utilization of processes: in particular, *control groups* and *name space isolation*. These features allow for the layering of new private virtual file system components on top of the host file system and a special partitioned process space for applications to run using libraries virtually stored in the layered file system. For all practical purposes, a program running in a container looks like it is running in its own VM, but without the extra baggage of a complete OS. A contained application uses the resources of the host OS, which can even control the amount of resource devoted to each container: for example, the CPU percentage and the amount of memory and disk space.

By mid-2013, a little company called dotCloud released a tool that provided a better way to deploy encapsulated applications. This tool became Docker [71], and dotCloud became Docker, Inc. `docker.com`. Microsoft also figured out how to do the same thing with Windows. While other container technologies exist, we focus primarily on Docker in this book.

Docker allows applications to be provisioned in containers that encapsulate all application dependencies. The application sees a complete, private process space, file system, and network interface isolated from applications in other containers on the same host operating system. Docker isolation provides a way to factor large applications, as well as simple ways for running containers to communicate with each other. When Docker is installed on Linux, Windows 10, or Mac, it runs on a base Linux kernel called Alpine that is used for every container instance. As we describe below, additional OS features are layered on top of that base. This layering is the key to container portability across clouds.

Docker is designed to support a variety of distributed applications. It is now widely used in the Internet industry, including by major companies like Yelp, Spotify, Baidu, Yandex, and eBay. Importantly, Docker is supported by major public cloud providers, including Google, Microsoft, Amazon, and IBM.

To understand how containers are built and used, one must understand how the file system in a container is layered on top of the existing host services. The key is the **Union File System** (more precisely, the advanced multilayered unification file system (AuFS) and a special property called *copy on write* that allows the system to reuse many data objects in multiple containers.

Docker images are composed of layers in the Union File System. The image is itself a stack of read-only directories. The base is a simplified Linux or Windows file system. As illustrated in figure 6.1, additional tools that the container needs are then layered on top of that base, each in its own layer. When the container is run, a final writable file system is layered on top.

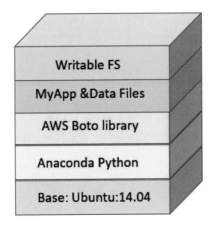

Figure 6.1: The Docker Union File System is layered on a standard base.

As an application in the container executes, it uses the writable layer. If it needs to modify an object in the read-only layers, it copies those objects into the writable layer. Otherwise, it uses the data in the read-only layer, which is shared with other container instances. Thus, typically only a little of the container image needs to be actually loaded when a container is run, which means that containers can load and run much faster than virtual machines. In fact, launching a container typically takes less than a second, while starting a virtual machine can take minutes.

In addition to the file system layers in the container image, you can mount a host machine directory as a file system in the container's OS. In this way a container can share data with the host. Multiple containers can also share these mounted directories and can use them for basic communication of shared data.

6.2   Docker and the Hub

The Docker website, Docker.com, provides the tools needed to install Docker on a Linux, Mac, or Windows 10 PC. This site also links to the Docker Hub, a public resource where you can store your own containers, and search for and download any of the hundreds of public containers.

Install Jupyter with Docker on your laptop. You must first install Docker on your machine. While the details differ on Linux, Mac, or PC, the installation is a simple process, similar to that of installing a new browser or other desktop application. Follow the download and install instructions on the Docker.com website. Docker does not have a graphical interface: it is based on a command line API. Hence, you need to open a "powershell" or "terminal" window on your machine. The Docker commands are then the same on Linux, Mac, or PC.

Once you have installed Docker, you can verify that it is running by executing the docker ps command, which tells you which containers are running. You should see the following output, as no containers are running.

```
C:\> docker ps
CONTAINER ID   IMAGE   COMMAND   CREATED   STATUS   PORTS      NAMES
C:\>
```

We launch Jupyter with the docker run command. The following example uses two of the many parameters supported. The first flag, -it, causes the printing of a URL with a token that you can use to connect to the new Jupyter instance. The second flag, -p 8888:8888, binds port 8888 in the container's IP stack to port 8888 on our machine. Finally, the command specifies the name of the container, jupyter/scipy-notebook, as it can be found in the Docker Hub.

```
C:\> docker run -it -p 8888:8888 jupyter/scipy-notebook
Copy/paste this URL into your browser when you connect for the
first time, to login with a token:
  http://localhost:8888/?token=b9fc19aa8762a6c781308bb8dae27a...
```

Rerunning the docker ps command shows that our newly started Jupyter notebook is now running. (Confusingly, each of the two output lines is wrapped.)

```
C:\> docker ps
CONTAINER ID   IMAGE                   COMMAND              CREATED
STATUS         PORTS                   NAMES
6cb4532fa0b    jpyter/scipy-notebook   "tini--start-note"   6 seconds ago
up 5 seconds    0.0.0.:8888->8888/tcp  prickly_meitner
C:\>
```

The first time you execute this command for a specific container, it must search for and download various elements of the container file system, which may take several minutes. Then, because the container jupyter/scipy-notebook is in the Docker Hub, it finds the container there and begins to download it. Once completed, it then starts the container. The container image is now local; thus, the next time that you run it, it can start in a few seconds.

The docker ps output includes an autogenerated instance name, in this case prickly_meitner. To kill the instance, run docker kill prickly_meitner.

There are a number of standard Docker features that are good to know. The flag -it connects the container's standard I/O to the shell that ran the docker command. We used that when starting Jupyter, which is why we saw the output. If you do not want to interact with the container while it is running, you can use the flag -d to make it run in detached mode.

Also useful is the Docker mechanism that allows a container to access a disk volume on the host machine, so that a process in your container can save files (when a container is terminated, its file system goes away as well), or access your local data collections. To mount a local directory on your laptop as a volume on the Docker container file system, use the -v localdir:/containername flag. (If you are running on Windows 10, you need to access the Docker settings and give Docker permission to see and modify drive C.)

The following commands illustrates the use of both -it and -v. We first use the docker command on a Mac to launch a Linux Ubuntu container with the Mac's /tmp directory mounted as /localtmp. Due to -it, we are presented with a command prompt for the newly started Ubuntu container. We then run df in the container to list its file systems, which include /localtmp.

```
docker run -it -v /tmp:/localtmp ubuntu
root@3148dd31e6c7:/# df
Filesystem      1K-blocks      Used Available Use% Mounted on
none             61890340  41968556  16754860  72% /
tmpfs             1022920         0   1022920   0% /dev
tmpfs             1022920         0   1022920   0% /sys/fs/cgroup
osxfs          975568896 143623524 831689372  15% /localtmp
/dev/vda2        61890340  41968556  16754860  72% /etc/hosts
shm                65536         0     65536   0% /dev/shm
root@3148dd31e6c7:/#
```

Notice that when we connect to the new container, we do so as root. Running Jupyter always presents a security challenge, especially if you are running it on a machine with a public IP address. You should certainly use HTTPS and a password, especially if you are running it on a remote VM in the cloud. We can configure these options by using the -e flag on the run command to pass environment flags through to Jupyter. For example, -e GEN_CERT=yes tells Jupyter to generate a self-signed SSL certificate and to use HTTPS instead of HTTP for access. To tell Jupyter to use a password, we need to do a bit more work. Start Python and issue the following commands to created a hashed password:

```
In [1]: import IPython
In [2]: IPython.lib.passwd()
Enter password:
Verify password:
Out[2]: 'sha1:db02b6ac4747:fc0561c714e52f9200a058b529376bc1c7cb7398'
```

Remember your password and copy the output string. Let's assume that we also want to mount a local directory `c:/tmp/docmnt` as a local directory `docmnt` inside the container. Jupyter has a user called `jovyan` and the working directory is `/home/jovyan/work`. The complete command for running Jupyter is then:

```
$ docker run -e GEN_CERT=yes -d -p 8888:8888 \
    -v /tmp/docmnt:/home/jovyan/work/docmnt \
    jupyter/scipy-notebook start-notebook.sh \
    --NotebookApp.password='sha1:.... value from above'
```

This command launches Jupyter via HTTPS with your new password. When the container is up, you can connect to it via HTTPS at your host's IP address and port 8888. Your browser may complain that this is not a valid web page, because you have created a self-signed certificate for this site. You can accept the risk, and you should see the page shown in figure 6.2.

Figure 6.2: View of Jupyter in the browser after accepting the security exception.

We'll mention one last flag: If you are running on a machine with multiple cores, you can use `-cpuset-cpus` to specify how many cores to use for your container. For example, `-cpuset-cpus 0-7` specifies that cores 0 through 7 are to be used.

6.3 Containers for Science

An impressive number of scientific applications have already been containerized, and the number is growing daily. The following are just a few of those to be found in the Docker repository.

- Radio astronomy tools are available, including containers for LOFAR, PyImager, and MeqTree.

- For bioinformatics, the ever-popular Galaxy toolkit is available in various forms. The University of Hamburg genome toolkit is also available.

- For mathematics and statistics, there are R and Python, which have been packaged with NumPy; others also are available in various combinations.

- For machine learning, there are complete collections of ML algorithms written in Julia as well as many versions of Spark, the Vowpal Wabbit tools, and the scikit-learn Python tools.

- For geospatial data, there is a container with geoserver.

- For digital archiving and data curation, there are containers for DSpace and iRODS.

- The iPlant consortium has developed the Agave science-as-a-service platform `agaveapi.co`, the various components of which are now containerized.

- The UberCloud project `theubercloud.com` has produced many containerized science and engineering applications.

In each case, you can spin up a running instance of the software in seconds on a Linux, Mac, or PC with Docker installed. Can we assume, then, that all the problems of science and engineering in the cloud are solved? Unfortunately not. What if you want to run a cluster of Docker containers that are to share a large workload? Or a big Spark deployment? We cover these topics in chapter 7.

Binder represents another excellent use of containers. This tool allows you to take a Jupyter notebook in GitHub and automatically build a container that, when invoked from GitHub, is launched on a Kubernetes cluster. (We describe Kubernetes in section 7.6.5 on page 120.)

6.4   Creating Your Own Container

Creating your own container image and storing it in the Docker Hub is simple. Suppose you have a Python application that opens a web server, waits for you to provide input, and then uses that input to pull up an image and display it. Now, let's build this little server and its image data as a container. We use a Python application based on the Bottle framework for creating the web server. (All code for this example is available in the **EXTRAS** tab of the book website.) Assume the images are all stored as `jpg` files in a directory called `images`. We are going to use this container as the basis for other scientific Python containers, so we make sure that it includes all the standard SciPy tools. In addition, since future versions of this container will want to interact with Amazon, we include the Amazon Boto3

SDK in the container. We now have all the elements we need for the container's layered file system. We next create a file named `Dockerfile`, as follows.

```
FROM jupyter/scipy-notebook
MAINTAINER your name <yourname@gmail.com>
RUN pip install bottle
COPY images /images
COPY bottleserver.py /
ENTRYPOINT ["ipython", "/bottleserver.py" ]
```

The first line specifies that we want to build on `jupyter/scipy-notebook`, a well-maintained container in the Docker Hub. We could have started at a lower level, such as basic Ubuntu Linux. But `jupyter/scipy-notebook` has everything we need except for Boto3 and Bottle, so we add those layers by running a `pip install` for each. We next add our images and finally our Python source code. The `ENTRYPOINT` line tells Docker what to execute when the container runs. We can now build the container with the `docker build` command. The result is shown in figure 6.3 on the following page.

The first time you run `docker build`, it downloads all the components for `jupyter/scipy-notebook`, Boto3, and Bottle, which takes a minute or so. After these components have been downloaded, they are cached on your local machine. Note that all the `pip install`s have been run and layered into the file system. Even the Python code was parsed to check for errors (see step 7). Because of this preinstallation, when the container is run, everything is there already.

```
Docker run -d -p 8000:8000 yourname/bottlesamp
```

Create a free Docker account and save your container to the Docker Hub as follows. You can then download your container to the cloud and run it there.

```
docker push yourname/bottlesamp
```

6.5 Summary

Containers provide an excellent tool for encapsulating scientific applications in a way that allows them to be deployed on any cloud. We have shown how to deploy Jupyter on any Docker-enabled laptop or cloud VM with one `docker run` command. Many other scientific applications have also been packaged as containers and are easily downloaded and used, from standard bioinformatics tools such as Galaxy to commercial packages such as Matlab.

```
> docker build -t="yourname/bottlesamp" .
Sending build context to Docker daemon 264.2 kB
Step 1 : FROM jupyter/scipy-notebook
 ---> 3e6809ce29ee
Step 2 : MAINTAINER Your Name <you@gmail.com>
 ---> Using cache
 ---> 5f09a5508c7b
Step 3 : RUN pip install boto3
 ---> Using cache
 ---> 74cfec535986
Step 4 : RUN pip install bottle
 ---> Using cache
 ---> d5a33c1b900a
Step 5 : COPY images /images
 ---> Using cache
 ---> 8cff8cd7c147
Step 6 : COPY bottleserver.py /
 ---> ebcb834dcc23
Removing intermediate container 8b3415f5ab12
Step 7 : ENTRYPOINT ipython /bottleserver.py
 ---> Running in c6d63ce5b327
 ---> c0ba12fd36d8
Removing intermediate container c6d63ce5b327
Successfully built c0ba12fd36d8
```

Figure 6.3: Output from a Docker build command.

We have also shown how to build a container from the most basic layer, layering your own applications and their dependencies on top. Containers can run on a fraction of a core or control multiple cores on a multicore server. Containers can mount directories in the host system as volumes, and multiple containers on a single host can share these volumes. A tool called Docker-compose, not discussed here, allows containers to communicate with each other by messages. We describe in chapter 7 how to coordinate hundreds of containers on large clusters.

6.6 Resources

We have covered only a few Docker capabilities here. Fortunately, excellent online resources are available. Of the many books on Docker, we particularly like *The Docker Book* [249] and the more recent *Docker: Up & Running* [192]. Other technologies exist for building and executing containers. Singularity [177] `singularity.lbl.gov` has attracted interest from the HPC community.

Chapter 7

Scaling Deployments

"I fixed my eyes on the largest cloud, as if, when it passed out of my sight, I might have the good luck to pass with it."

—Sylvia Plath, *The Bell Jar*

Public clouds are built from the ground up to allow customers to scale their deployed services to fit their needs. The most common motivation for scaling in industry is to allow services to support thousands or millions of concurrent users. For example, an online media streaming site may initially be deployed on a single VM instance. As its business grows, it may need to scale to use 100s or even 10,000s of servers at peak times, and then scale back when business is slow.

As we discuss in chapter 14, such multiuser scaling can also be of interest to some research users. However, the more urgent need facing scientists and engineers is often to run a bigger simulation or to process more data. To do so, they need to be able to obtain access to 10 or 100 or 1,000 servers quickly and easily, and then run parallel programs efficiently across those servers. In this chapter, we describe how different forms of parallel computation can be mapped to the cloud, and some of the many software tools that clouds make available for this purpose.

We return to various of these tools in part III of the book, when we look at data analytics, streaming, and machine learning, each of which increasingly require large-scale computing.

7.1 Paradigms of Parallel Computing in the Cloud

We consider in this chapter five commonly used parallel computing paradigms for scaling deployments in the cloud.

The first is highly synchronous **single program multiple data** (SPMD) computing. This paradigm dominates supercomputer applications and is often the first that scientific programmers want to try when they meet the cloud. We will show how to use SPMD programming models in the cloud, pointing out where care must be taken to get good performance.

The second paradigm is **many task parallelism**, in which a large queue or queues of tasks may be executed in any order, with the results of each task stored in a database or in files, or used to define a new task that is added to a queue. Such computations were termed *embarrassingly* parallel by early parallel computing researchers [131]; we prefer *happily* parallel. It adapts well to clouds.

The third paradigm is **bulk synchronous parallelism** (BSP) [139]. Originally proposed as a model for analyzing parallel algorithms, BSP is based on processes or threads of execution that repeatedly perform independent computations, exchange data, and then synchronize at a barrier. (When a process reaches a barrier in its program, it stops and cannot proceed until all other threads/processes in the synchronization group reach the same point.) A contemporary example of the BSP model is the **MapReduce** style made famous by Google and now widely used in its **Hadoop** and **Spark** realizations.

The fourth paradigm is the **graph execution model**, in which computation is represented by a directed, usually acyclic, graph of tasks. Execution begins at a source of the graph. Each node is scheduled for execution when all incoming edges to that node come from task nodes that have completed. Graphs can be constructed by hand or alternatively generated by a compiler from a more traditional-looking program that describes the graph either implicitly or explicitly. We describe graph execution systems in chapters 8 and 9, including the data analytics tool **Spark** and the **Spark Streaming**, **Apache Flink**, **Storm**, and **Google Dataflow** systems. The graph execution model is also used in machine learning tools, such as the **Google TensorFlow** and **Microsoft Cognitive Toolkit** systems that we discuss in chapter 10. In many of these cases, the execution of a graph node involves a parallel operation on one or more distributed data structures. If the graph is just a linear sequence of nodes, then execution becomes an instance of BSP concurrency.

The fifth paradigm we consider is **microservices** and actors. In the actor model of parallel programming, computing is performed by many **actors** that communicate via messages. Each actor has its own internal private memory and goes into action when it receives a message. Based on the message, it can change its internal state and then send messages to other actors. You can implement an actor as a simple web service. If you limit the service to being largely stateless, you are now in the realm of **microservices**, a dominant design paradigm for large cloud applications. Microservices are often implemented as container instances, and they depend on robust container management infrastructure such as **Mesos** [154] and **Kubernetes** [79], which we describe in later sections.

We discuss in the following how each of these paradigms can be implemented in cloud environments. We look at the impact that GPU accelerators are having on science in the cloud and examine how we can build traditional high-performance computing (HPC) clusters. We then move on to more cloud-centric systems that have evolved for scalable data analytics. This last group includes Mesos and Kubernetes for managing large collections of microservices, as well as MapReduce-style computing systems such as Hadoop and Spark.

7.2 SPMD and HPC-style Parallelism

Scientists and engineers frequently ask, "Can I do traditional high-performance computing (HPC) in the cloud?" The answer used to be, "not really": Indeed, a 2011 U.S. Department of Energy report [223] and a 2012 National Aeronautics and Space Administration (NASA) report [196] both concluded that cloud did not meet expectations for HPC applications. However, things have changed as cloud providers have deployed both high-speed node types and specialized interprocessor communication hardware. It is now feasible to achieve excellent performance for HPC applications on cloud systems for modest numbers of processing cores, as we now discuss.

7.2.1 Message Passing Interface in the Cloud

The Message Passing Interface (MPI) remains the standard for large-scale parallel computing in science and engineering. The specification allows for portable and modular parallel programs; implementations allow communicating processes to exchange messages frequently and at high speed. Amazon `aws.amazon.com/hpc` and Azure `azure.microsoft.com/en-us/solutions/big-compute` both offer special HPC-style cluster hardware in their data centers that can achieve high performance for MPI applications. Cycle Computing reported back in 2015 that "we have seen comparable performance [for MPI applications on EC2 vs. supercomputers] up to 256 cores for most workloads, and even up to 1024 for certain workloads that aren't as tightly-coupled." [156] The key to getting good performance is configuring advanced networking, as we discuss below when we describe how to create virtual clusters on Amazon and Azure.

7.2.2 GPUs in the Cloud

Graphics processing units have been used as accelerators in supercomputing for about 10 years. Many top supercomputing applications now rely on compute nodes with GPUs to give massive accelerations to dense linear algebra calculations. Coincidentally this type of calculation is central to the computations needed to train **deep neural networks** (NNs), which lie at the nexus of interest in machine learning by the technology industry. Consequently it is no surprise that GPUs are making their way into cloud servers.

The most common GPUs used in the current systems are from NVIDIA and are programmed with a single instruction multiple data (SIMD) model in a system called CUDA. SIMD operators are, in effect, array operations where each instruction applies to entire arrays of data. For example, a state-of-the-art system in 2017, the NVIDIA K80, has 4,992 CUDA cores with a dual-GPU design delivering up to 2.91 teraflops (10^{12} floating point operations per second) double-precision performance and 8.73 teraflops in single precision. Amazon, Microsoft, and IBM have made K80-supported servers available in their public clouds. The Amazon EC2 P2 instances incorporate up to eight NVIDIA Tesla K80 accelerators, each running a pair of NVIDIA GK210 GPUs. The Azure NC24r series has four K80s with 224 GB memory, 1.5 TB solid state disk (SSD), and high-speed networking. You do not need to configure a particularly large cluster of these machine instances to create a petaflop (10^{15} floating point operations/second) virtual computer.

> **MPI cloud computing for proton therapy**. This example illustrates how specialized node types (in this case, GPU-equipped nodes with 10 gigabit/s interconnect) allow cloud computing to deliver transformational computing power for a time-sensitive medical application. It also involves the use of MPI for inter-instance communication, Apache Mesos for acquiring and configuring virtual clusters, and Globus for data movement between hospitals and cloud.
>
> Chard and colleagues have used cloud computing to reconstruct three-dimensional proton computed tomography (pCT) images in support of proton cancer treatment [91]: see figure 7.1 on the following page. In a typical clinical usage modality, protons are first used in an online position verification scan and are then targeted at cancerous cells. The results of the verification scan must be returned rapidly.

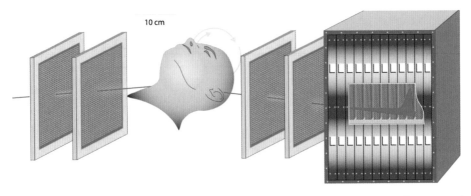

Figure 7.1: Proton computed tomography. Protons pass left to right through sensor planes and traverse the target before stopping in the detector at the far right.

A single reconstruction can require the analysis of around two billion 50-byte proton histories, resulting in an input dataset of ∼100 GB. The reconstruction process is complex, involving multiple processes and multiple stages. First, each participating process reads a subset of proton histories into memory and performs some preliminary calculations to remove abnormal histories. Then, filtered back projection is used to estimate an initial reconstruction solution, from which most likely paths (MLPs) are estimated for the proton through the target. The voxels of the MLP for each proton identify the nonzero coefficients in a set of nonlinear equations that must then be iteratively solved to construct the image. This solution phase takes the bulk of the time and can be accelerated by using GPUs and by caching the MLP paths (up to 2 TB for a 2-billion-history image) to avoid recomputation.

Chard et al. use a high-performance, MPI-based parallel reconstruction code developed by Karonis and colleagues that, when run on a standalone cluster with 60 GPU-equipped compute nodes, can reconstruct two billion histories in seven minutes [167]. They configure this code to run on an Amazon virtual cluster of GPU-equipped compute instances, selected to meet requirements for compute performance and memory. They deploy on each instance a VM image configured to run the pCT software plus associated dependencies (e.g., MPI for inter-instance communication).

The motivation for developing this pCT reconstruction service is to enable rapid turnaround processing for pCT data from many hospitals, as required for clinical use of proton therapy. Thus, the pCT service also incorporates a scheduler component, responsible for accepting requests from client hospitals, acquiring and configuring appropriately sized virtual clusters (the size depending on the number of proton histories to be processed), and dispatching reconstruction tasks to those clusters. They use the Apache Mesos scheduler (section 7.6.6 on page 124) for this task and the Globus transfer service (section 3.6 on page 51) for data movement.

This pCT service can compute reconstructions for $10 per image when using

Amazon spot instances (see page 80 for a discussion of spot instances). This is a revolutionary new capability, considering that the alternative is for each hospital with a proton therapy system to acquire, install, and operate a dedicated HPC cluster.

7.2.3   Deploying an HPC Cluster on Amazon

A useful tool for building HPC clusters on the Amazon cloud is Amazon's **Cloud-Formation** service, which enables the automated deployment of complex collections of related services, such as multiple EC2 instances, load balancers, special subnetworks connecting these components, and security groups that apply across the collection. A specific deployment is described by a template that defines all needed parameters. Once required parameter values are set in the template, you can launch multiple identical instances of the described deployment. You can both create new templates and modify existing templates.

We do not go into those complexities here but instead consider one particular case: **CfnCluster (CloudFormation Cluster)**. This tool is a set of Python scripts that you can install and run on your Linux, Mac, or Windows computer to invoke CloudFormation, as follows, to build a private, custom HPC cluster.

```
sudo pip install cfncluster
cfncluster configure
```

The configuration script asks you various questions, including the following.

1. A name for your cluster template (e.g., "mycluster")

2. Your Amazon access keys

3. The region in which you want your cluster to be deployed

4. The name for your virtual private cloud (VPC)

5. The name for the key pair that you want to use

6. Virtual Private Cloud (VPC) and subnet IDs

The following command causes a basic cluster to be launched, in the manner illustrated in figure 7.2 on the following page. (As you have not yet specified the number or type of compute nodes to start, or what image to run on those nodes, defaults are used.) This process typically takes about 10 minutes.

```
cfncluster create mycluster
```

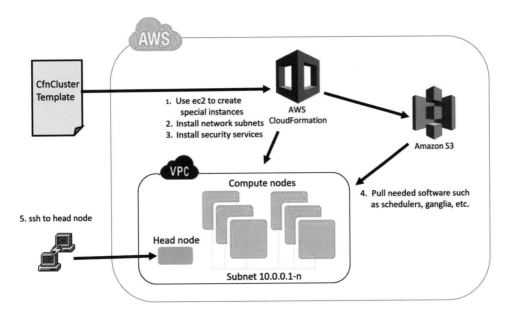

Figure 7.2: CloudFormation steps involved in launching a private HPC cloud from a CfnCluster template.

The "create" command returns a **Ganglia** URL. Ganglia is a well-known and frequently used cluster monitoring tool. Following that link takes you to a Ganglia view of your HPC cluster. The default settings for a new cluster are *autoscale* compute nodes and the *gridEngine* scheduler. This combination works well if you are running a lot of non-MPI batch jobs. With autoscale, compute nodes are shut down when not in use, and new nodes are started when load increases.

Now let us deploy a new cluster with better compute nodes and a better scheduler. The following command deletes your current deployment.

```
cfncluster delete mycluster
```

To create a new and improved deployment that you can use for MPI programs, find the directory ~/.cfncluster on your PC and edit the file config. Look for a section called [cluster mycloud], and add or edit the following four lines in that section.

```
compute_instance_type = c3.xlarge
initial_queue_size = 4
maintain_initial_size = true
scheduler = slurm
```

The choice of compute instance type here is important. The `c3.xlarge` instance type supports what Amazon calls **enhanced networking**, which means that it runs on hardware and with software that support Single-Root I/O Virtualization (SR-IOV). As we discuss in more detail in section 13.3.2 on page 286, this technology reduces latency and increases bandwidth by allowing VMs to access the physical network interface card (NIC) directly.

You do not need to specify a VM image because the default contains all libraries needed for HPC MPI-style computing. You have also stated that you want all of your compute nodes to stay around and not be managed by autoscale, and that you want *Slurm* to be the scheduler. We have found that these options make interactive MPI-based computing much easier. When you now issue the create command, as follows, CloudFormation deploys an HPC cluster with these properties.

```
cfncluster create mycluster
```

Now let us run an MPI program on the new cluster. You first login to the cluster's head node, using the key pair that we used to create the cluster. On a PC you can use PuTTY, and on a Mac you can use `ssh` from the command line. The user is `ec2-user`. First you need to set up some path information. At the command prompt of the head node, issue the following three commands.

```
export PATH=/usr/lib64/mpich/bin:$PATH
export LD_LIBRARY_PATH=/usr/lib64/mpich/lib
export I_MPI_PMI_LIBRARY=/opt/slurm/lib/libpmi.so
```

You next need to know the local IP addresses of your compute nodes. Create a file called `ip-print.c`, and put the following C code in it.

```c
#include <stdio.h>
#include <mpi.h>
#include <stdlib.h>

main(int argc, char **argv)
{
    char hostname[1024];
    gethostname(hostname, 1024);
    printf("%s\n", hostname);
}
```

You can then use the Slurm command **srun** to run copies of this program across the entire cluster. For example, if your cluster has 16 nodes, run:

```
mpicc ip-print.c
srun -n 16 /home/ec2-user/a.out > machines
```

The output file **machines** should then contain multiple IP addresses of the form **10.0.1.x**, where **x** is a number between 1 and 255, one for each of your compute nodes. (Another way to find these private IP addresses is to go to the EC2 console and look at the details for each running server labeled *Compute*. You need to do this for each instance, so if you have a lot of instances, the above method is faster.)

Now you can run the simple MPI program in figure 7.3 on the next page to test the cluster. Call this **ring_simple.c**. This program starts with MPI node 0 and sends the number -1 to MPI node 1. MPI node 1 sends 0 to node 2, node 2 sends 1 to node 3, and so on. Compile and run this with the command below.

```
mpicc ring.c
mpirun -np 7  -machinefile ./machines /home/ec2-user/a.out
```

Assuming seven processes running on four nodes, you should get results like the following.

```
Warning: Permanently added the RSA host key for IP address '10.0.1.77' to
the list of known hosts.
Warning: Permanently added the RSA host key for IP address '10.0.1.78' to
the list of known hosts.
Warning: Permanently added the RSA host key for IP address '10.0.1.76' to
the list of known hosts.
Warning: Permanently added the RSA host key for IP address' 10.0.1.75' to
the list of known hosts.
Received number -1 from process 0 on node ip-10-0-1-77
Received number 0 from process 1 on node ip-10-0-1-75
Received number 1 from process 2 on node ip-10-0-1-78
Received number 2 from process 3 on node ip-10-0-1-76
Received number 3 from process 4 on node ip-10-0-1-77
Received number 4 from process 5 on node ip-10-0-1-75
Received number 5 from process 6 on node ip-10-0-1-78
```

Observe that the **mpirun** command distributed the MPI nodes uniformly across the cluster. You are ready to do some real HPC computing on your virtual private MPI cluster. You might, for example, measure message transit times. We converted our C program into one in which our nodes are connected in a ring, and measured the time for messages to go around this ring ten million times. We found that the average time to send a message and have it received at the destination was about 70 microseconds. While this is not a standard way to measure message latencies, the result is consistent with that seen in other cloud cluster implementations. It is at least 10 times slower than conventional supercomputers, but note that the underlying protocol used here is IP and that we do not use any special cluster network such as InfiniBand. The full code is available in GitHub and is linked from the book website in the **EXTRAS** tab.

```
#include <mpi.h>
#include <stdio.h>
#include <stdlib.h>
#include <time.h>

int main(int argc, char** argv) {
  // Initialize the MPI environment
  MPI_Init(NULL, NULL);
  // Find out rank, size
  int rank, world_size, number;
  MPI_Comm_rank(MPI_COMM_WORLD, &rank);
  MPI_Comm_size(MPI_COMM_WORLD, &world_size);
  char hostname[1024];
  gethostname(hostname, 1024);

  // We assume at least two processes for this task
  if (world_size < 2) {
    fprintf(stderr, "World size must be >1 for %s\n", argv[0]);
    MPI_Abort(MPI_COMM_WORLD, 1);
  }

  if (rank == 0) {
    // If we are rank 0, set number to -1 & send it to process 1
    number = -1;
    MPI_Send(&number, 1, MPI_INT, 1, 0, MPI_COMM_WORLD);
  }
  else if (rank > 0 && rank < world_size) {
    MPI_Recv(&number, 1, MPI_INT, rank-1, 0, MPI_COMM_WORLD,
             MPI_STATUS_IGNORE);
    printf("Received number %d from process %d on node %s\n",
           number, rank-1, hostname);
    number = number+1;
    if (rank+1 < world_size)
      MPI_Send(&number, 1, MPI_INT, rank+1, 0, MPI_COMM_WORLD);
  }
  MPI_Finalize();
}
```

Figure 7.3: The MPI program `ring_simple.c` that we use for performance experiments.

7.2.4 Deploying an HPC Cluster on Azure

We describe two approaches to building an HPC cluster on Azure. The first is to use Azure's service deployment orchestration service, **Quick Start**. Like Amazon CloudFormation, it is based on templates. The templates are stored in GitHub and can be invoked directly from the GitHub page. For example, figure 7.4 shows the template start page for creating a Slurm cluster.

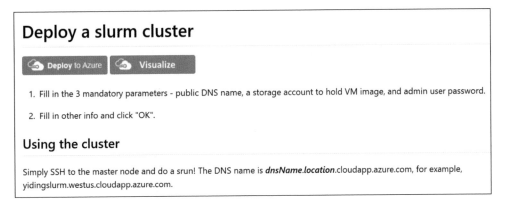

Figure 7.4: Azure Slurm start page [8].

When you click Deploy to Azure, you are taken to the Azure login page and then directly to the web form for completing the Slurm cluster deployment. At this point you enter the names for the new *resource group* that defines your cluster, the number and type of compute nodes, and a few details about the network. We refer you to the Microsoft tutorial [200] for more details on how to build a Slurm cluster and launch jobs on that cluster.

A second approach to HPC computing on Azure is **Azure Batch**, which supports the management of large pools of VMs that can handle large batch jobs, such as many task-parallel jobs (described in the next section) or large MPI computations. As illustrated in figure 7.5 on the next page, four major steps are involved in using this service.

1. Upload your application binaries and input data to Azure storage.

2. Define the pool of compute VMs that you want to use, specifying the desired VM size and OS image.

3. Define a *Job*, a container for tasks executed in your VM pool.

4. Create tasks that are loaded into the Job and executed in the VM pool.

Tasks are usually just command line scripts to be executed on each node. Typically the first tasks are executed on all nodes and involve loading your input data and applications, and performing other needed initialization of the individual VMs. The next task can be your MPI job. You can specify that this task not start until the previous task finishes. The final task can be one that moves your output data back to the Azure blob storage.

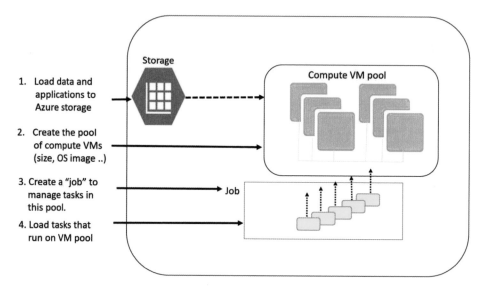

Figure 7.5: Steps in creating and executing an Azure batch job.

7.2.5 Scaling Further

We indicated above that tightly coupled HPC applications can scale to modest-sized clusters. Why can't they scale further? There are good reasons. Supercomputer architectures are designed with a heavy emphasis on the performance of the processor interconnection network. They use specialized hardware features and communication protocols to achieve message latencies in the sub-microsecond range. On the other hand, cloud data centers were originally built with networks based on the standard Internet TCP/IP protocols over Ethernet switches, with an emphasis on bandwidth from the cluster to the outside user rather than between servers.

A quantity called the network **bisection bandwidth** is a measure of how much data traffic can flow from one half of the supercomputer to the other half in a specified period of time. The networks used in supercomputers have an extremely high bisection bandwidth. The first cloud data centers had low bisection bandwidth. More recently, most public cloud vendors have redesigned their networks to take advantage of software defined networking and better network topologies, with the result that Google claims today to have a bisection bandwidth of more than 1 petabit/s (10^{15} b/s)—which, as they point out, is "enough for 100,000 servers to exchange information at 10 Gb/s each" [250]. In addition, the HPC subclusters deployed in Microsoft and Amazon data centers are now supplemented by more sophisticated InfiniBand as well as custom networks. Thus, it is becoming increasingly feasible to run latency-sensitive applications on clouds.

Nevertheless, one needs to consider the nature of the **service level agreement** (SLA) that cloud providers make with their users. Supercomputers commit to a specific processor type, network bandwidth, bisection width, and latency, allowing the user to predict an application's performance profile with a fair degree of certainty. With cloud data centers, however, the specific resources used to constitute a HPC subcluster in the cloud are less transparent. In other words, the SLA is complex and less likely to scale to large deployments than a supercomputer.

7.3 Many Task Parallelism

Suppose you need to analyze many data samples. Each analysis task can be performed independently of all the other tasks. This style of computation is simple. You place all data samples in a queue in the cloud, and then start a large number of worker VMs or containers. We refer to this as **many task parallelism**, but it is also known as **bag of tasks parallelism** and **manager worker parallelism**. Each worker repeatedly pulls a sample from a queue, processes the data, and stores the result in a shared table: see figure 7.6.

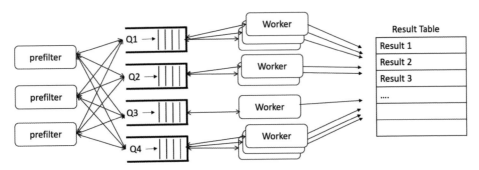

Figure 7.6: Simple many task execution model.

A good example application is genomics, in which many DNA sequences must often be processed. Wilkening et al. have analyzed the suitability of the cloud for this task [260]. Seattle Children's Hospital used the same approach for the NCBI BLAST code on Azure [33]. The cloud-based Globus Genomics, described in detail in section 14.4 on page 303, provides many genomics pipelines [187]. We present a detailed implementation of a many task example later in this chapter.

This strategy requires an efficient mechanism that multiple workers can each use to pull a unique task from the queue and push an item into a table. We also need a way to start and stop tasks. We describe below a simple Python framework for many task parallelism that uses a resource manager to coordinate worker activities.

7.4 MapReduce and Bulk Synchronous Parallelism

A slightly more sophisticated approach to parallelism is based on a concept called bulk-synchronous parallelism (BSP). This is important when worker tasks must periodically synchronize and exchange data with each other. The point of synchronization in a BSP computation is called a *barrier*, because no computation is allowed to proceed until all computations reach the synchronization point.

MapReduce is a special case of BSP computing. The concept is simple. Say you have a sequence of data objects X_i for $i = 1..n$ and you want to apply a function $f(x)$ to each element. Assume the result is a value in an associative ring like the real numbers, for which we can compose objects, and we want to compute the sum $\sum_{i=1}^{n} f(x_i)$. We *map f* over the data and then *reduce* by performing the sum. While this concept dates from Lisp programming in the 1960s, it was popularized for big data analysis in a 2004 paper by Google engineers Dean and Ghemawat [108] and became ubiquitous when Yahoo! released an open source implementation of MapReduce known as Hadoop.

A MapReduce computation starts with a distributed data collection partitioned into non-overlapping blocks. It then maps a supplied function over each block. Next it applies the reduction. This is where the barrier comes in. It combines the result of the map operator in a treelike way, perhaps through several stages, until it produces the result, as illustrated in figure 7.7 on the following page.

Even before Hadoop was released, people were applying MapReduce to scientific problems in biology [129, 246, 203], image analysis [188], and more [135]. Hadoop is a popular implementation of MapReduce that can run on any cloud and cluster. Hadoop is now part of the Apache YARN project, which we describe in chapter 8.

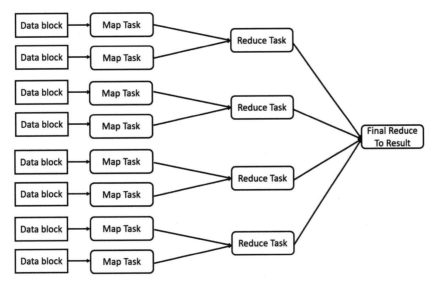

Figure 7.7: Simple MapReduce execution.

7.5 Graph Dataflow Execution and Spark

The MapReduce model can be thought of as the execution of a directed acyclic graph of tasks, but one can easily see how to generalize the idea. Define the computation as a graph of functions that are executed in a dataflow form until certain evaluation points are reached. Each function node in the graph represents a parallel invocation of the function on distributed data structures, as shown in figure 7.8 on the next page. At this level, the execution is essentially BSP-style, and the evaluation points are barrier synchronizations.

The dataflow graph is defined in a high-level control program that is then compiled into the parallel execution graph. Control flow starts with the control program; and when the graph is evaluated, control passes to a distributed execution engine. At the control points, the distributed data structures are converted back into structures that can be accessed by the control program. For example, if the algorithm involves an iteration, then the loop control is in the control program, and this is the barrier point for the parallel part of the program.

Spark `spark.apache.org` is a popular example of this style of computation. In Spark the control flow program is a version of SQL, Scala, or Python and, consequently, you can easily execute Spark programs from a Jupyter notebook. We provide many examples of such computations in this book. Spark was originally developed in the Berkeley AMPLab and is now supported by Databricks

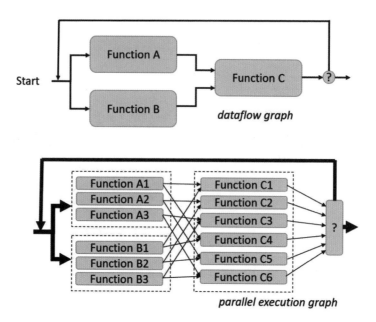

Figure 7.8: Dataflow graph as defined by the program (above) and after the parallelism is unrolled during execution (below).

`databricks.com`, among others. Databricks provides an online development environment and platform for deploying Spark-enabled clusters on Amazon.

Spark is also part of Microsoft's HDInsight toolkit and is supported on Amazon Elastic MapReduce. Microsoft documentation [21] describes how to deploy Spark on Azure with Linux or Windows and from a Jupyter notebook. Because Spark is used primarily for data analytics, we defer detailed examples to chapter 8.

7.6 Agents and Microservices

The cloud is designed to host applications that are realized as scalable services, such as a web server or the backend of a mobile app. Such applications accept connections from remote clients and, based on the client request, perform some computation and return a response. If the number of clients that are concurrently requesting service grows too large for one server to handle, the system should automatically spawn new instances of the server to share the load. Another scenario is an application that processes events from remote sensors, with the goal of informing a control system on how to respond, as when geosensors detecting ground motion tremors occurring in a significant pattern sound an earthquake warning. In this case the cloud part of the application may involve multiple components: sensor signal decoders, pattern analysis integrators, database searches, and alarm system interfaces. Another example is a large-scale data analytics system processing large amounts of text data: for example, a search index for the web.

This style of parallel programming is like an asynchronous swarm of communicating processes or services distributed over a virtual network in the cloud, as illustrated in figure 7.9. The individual processes may be *stateless*, such as a simple web service, or *stateful*, as in the actor programming model. Several great examples exist [56], but one that has been used in the public cloud is the Orleans systems from Microsoft Research [72].

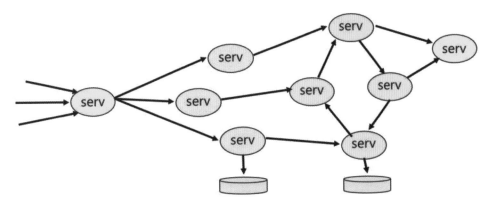

Figure 7.9: Conceptual view of a swarm of communicating microservices or actors.

Recently this asynchronous swarm style has been rebranded as a new paradigm, under the name **microservices**. The problem addressed by microservices is that of how to design and build a large, heterogeneous application that is secure, maintainable, fault tolerant, and scalable. This question is particularly important for large online services that need to support thousands of concurrent users. The app must remain up twenty-four hours a day while being maintained and upgraded. (This task is closely related to a software engineering methodology called **DevOps**, which integrates product delivery, quality testing, feature development, and maintenance releases in order to improve reliability and security and provide faster development and deployment cycles. From the programmer's perspective this is a "you built it, so now you run it" philosophy. Updates to parts of this system occur while the system is running, so the developers are tightly integrated with the IT professionals who manage the system.)

The microservice solution to this challenge is to partition the application into small, independent service components communicating with simple, lightweight mechanisms. The microservice paradigm design rules dictate that each microservice must be able to be managed, replicated, scaled, upgraded, and deployed independently of other microservices. Each microservice must have a single function and operate in a bounded context; that is, it has limited responsibility and limited dependence on other services. When possible one should reuse existing trusted services such as databases, caches, and directories. All microservices should be designed for constant failure and recovery. The communication mechanisms used by microservice systems are varied: they include REST web service calls, RPC mechanisms such as Google's Swift, and the Advanced Message Queuing Protocol (AMQP). The microservice philosophy has been adopted in various forms by many companies, including Netflix, Google, Microsoft, Spotify, and Amazon.

7.6.1 Microservices and Container Resource Managers

We next describe the basic ideas behind building microservices using clusters of containers. To build a substantial microservice application, you need to build on a platform that can manage large collections of distributed, communicating services. The microservices described below are instances of containers, and we use a resource manager to launch instances, stop them, create new versions, and scale their number up and down. Thanks to the open source movement and the public clouds, we have an excellent collection of tools for building and managing containerized microservices. We focus here on the Amazon ECS container service, Google Kubernetes, Apache Mesos, and Mesosphere on Azure.

7.6.2 Managing Identity in a Swarm

One issue that must be considered when you design an application around microservices is how you delegate your authority to a swarm of containerized, nearly stateless services. For example, if some microservices need to access a queue of events that you own, and others need to interact with a database that you created, then you need to pass part of your authority to those services so that they may make invocations on your behalf. We discussed this issue in section 3.2 on page 38, where we described how to manage your Amazon key pair.

One solution is to pass these values as runtime parameters through a secure channel to your remotely running application or microservice. This solution has two problems, however. First, microservices are designed to be shut down when not needed and scaled up in number when the demand is high. You would need to automate the process of fetching the credentials for each microservice reboot. Second, by passing these credentials, you endow your remote service with all of your authority. This is your microservice, so you are entitled to do so, but you may prefer to pass only a limited authority: for example, grant access only to the task queue but not the database.

The public cloud providers have a better solution: **role-based security**. What this means is that you can create special secure entities, called **roles**, that authorize individuals, applications, or services to access various cloud resources on your behalf. As we illustrate below, you can add a reference to a role in a container's deployment metadata so that each time the container is instantiated, the role is applied. (We discuss role-based access control in more detail in section 15.2 on page 319.)

7.6.3   A Simple Microservices Example

As in previous chapters, we present a single example and show how different resource managers can be used to implement that example. Several types of scientific applications can benefit from a microservice architecture. One common characteristic is that they are to run continuously and respond to external events. In chapter 9, we describe a detailed example of such an application: the analysis of events from online instruments and experiments. In this chapter, we consider the following simpler example.

> **Scientific document classification**. When scientists send technical papers to scientific journals, the abstracts of these papers often make their way onto the Internet as a stream of news items, to which one can subscribe via RSS feeds. A major source of high-quality streamed science data is arXiv `arxiv.org`, a collection of more than one million open-access documents. Other sources include the Public Library of Science (PLOS one), Science, and Nature, as well as online news sources. We have downloaded a small collection of records from arXiv, each containing a paper title, an abstract, and, for some, a scientific topic as determined by a curator. In the following sections we describe how to build a simple online science document classifier. Our goal is to build a system that pulls document abstracts from the various feeds and then uses one set of microservices to classify those abstracts into the major topics of physics, biology, math, finance, and computer science, and a second set to classify them into subtopic areas, as illustrated in figure 7.10 on the next page.
>
> The initial version that we describe here is more modest. In the first phase of this system, initial document classification, we feed the documents from a Jupyter notebook into a cloud-based message queue. The classifier microservices pull documents from the queue, perform the analysis, and push results into a NoSQL table, as shown in figure 7.11. This is now a simple many task system.

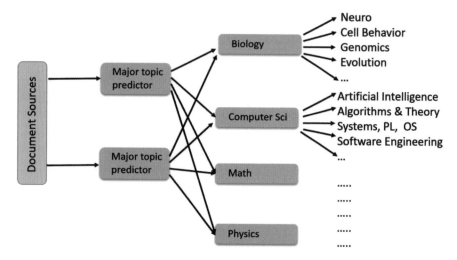

Figure 7.10: Online scientific document classifier example, showing two levels and subcategories for biology and computer science.

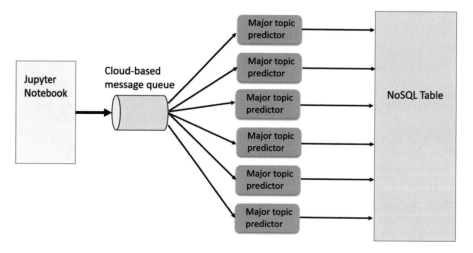

Figure 7.11: Document classifier version 1, showing the multiple predictor microservices.

7.6.4 Amazon EC2 Container Service

The Amazon **EC2 Container Service** (ECS) is a system to manage clusters of servers devoted to launching and managing microservices based on Docker containers. It comprises three basic components, as follows.

- One or more sets of EC2 instances, each a logical unit called a **cluster**. You always have at least one `Default` cluster, but you can add more.

- **Task definitions**, which specify information about the containers in your application, such as how many containers are part of your task, what resources they are to use, how they are linked, and which host ports they are to use.

- Amazon-hosted Docker image **repositories**. Storing your images here may make them faster to load when needed, but you can also use the public Docker Hub repository discussed in chapter 6.

First a note of potential confusion. Amazon refers to the EC2 VM instances in a cluster as *container instances*. These are *not* Docker containers. Later we add Docker containers to those VMs; but, using ECS terminology, those containers will be called tasks and services.

Before creating our cluster, you create two roles in the Amazon Identity and Access Management (IAM) system to address the identity management issues that we just discussed. The IAM link in the Security subarea of the AWS management console takes you to the IAM Dashboard. On the left, select `Roles`. From there, you can select `Create New Roles`. Name it `containerservice`, and then select the role type. You need two roles: one for the container service (which actually refers to the VMs in our cluster) and one for the actual Docker services that we deploy. Scroll down the list of role types and look for *Amazon EC2 Container Service Role* and *Amazon Container Service Task Role*. Select the Container Service Role for `containerservice`. Save this, and now create a second role and call it `mymicroservices`. This time, select the Amazon Container Service Task Role. Now when you go back to the dashboard, it should look like figure 7.12.

Figure 7.12: AWS IAM Console, showing the two new roles.

On the panel on the left is the link *Roles*. Select your `containerservice` role, and click (Roles). You should now be able to attach various access policies to your role. The attach policy button exposes a list of over 400 access policies that you can attach. Add three policies: *AmazonBigtableServiceFullAccess*, *AmazonEC2ContainerServiceforEC2role*, and *AmazonEC2ContainerServiceRole*.

Next, add two policies to the `mymicroservices` role, for the Amazon Simple Queue Service (SQS) and DynamoDB: *AmazonSQSFullAccess* and *AmazonDynamoDBFullAccess*. Finally, at the top of the page on which are listed the policies, you should see the *Role ARN* (Amazon Resource Name), looking something like `arn:aws:iam::01234567890123:role/mymicroservices`. Copy and save it.

Creating a cluster is now easy. From the Amazon ECS console, simply click (create cluster), and then give it a name. You next say what EC2 Instance type you want and provide the number of instances. If you need `ssh` access to the cluster nodes, include a cryptographic key pair; however, this is not needed for managing containers. You can use all the defaults on this page except that when you get to *Container Instance IAM role*, you should see the "containerservice" role. Select this, and select *create*. You should soon see the cluster listed on the cluster console, with the container instances running.

We next describe the steps needed to create a task definition and launch a service. We present the highlights here; full details are in notebook 9. We use the following example code to illustrate the approach. The call to `register_task_definition` creates a task definition in a family named `predict` with a standard network mode, and the *Role ARN* from above as `taskRoleArn`.

```
import boto3
client = boto3.client('ecs')
response = client.register_task_definition(
    family='predict',
    networkMode='bridge',
    taskRoleArn= 'arn:aws:iam::01233456789123:role/mymicroservices',
    containerDefinitions=[
        {
            'name': 'predict',
            'image': 'cloudbook/predict',
            'cpu': 20,
            'memoryReservation': 400,
            'essential': True,
        },
    ],
)
```

The rest of this code defines the container. It names the task definition `predict`, specifies that the image is `cloudbook/predict` from the Docker public Hub, and indicates a need for 20 units of computing (out of 1,024 available on a core) and 400 MB of memory. This is about as simple as a task definition can get. The notebook shows a more sophisticated task definition that manages port mappings.

Given a task definition, you can then call the `create_service` function, as in the following, which creates a service called `predictor` that is to run with eight microservices on the `cloudbook` cluster. Note the task definition name, `predict:5`. You often end up modifying various aspects of a task definition. Each time you execute the `register_task_definition` call, it creates a new revision of that task with a version tag. The suffix ":5" indicates that this is the fifth version.

```
response = client.create_service(
    cluster='cloudbook',
    serviceName='predictor',
    taskDefinition='predict:5',
    desiredCount=8,
    deploymentConfiguration={
        'maximumPercent': 100,
        'minimumHealthyPercent': 50
    }
)
```

Our invocation requires that at least 50% of the requested instances be granted. The first time we created this service, it took about a minute to download the 2 GB Docker image from the public Hub and load it into our cluster VMs. In subsequent runs, it took only a few seconds to start the service, because the image was local. If you now look at the cluster in the EC2 console, you see the state as shown in figure 7.13 on the next page.

Next we describe the `cloudbook/predict` service and how it works with the message queue. The Amazon SQS service is simple to use. Figure 7.14 shows a slightly abbreviated version of the prediction microservice. (Missing are the details of how to set up the DynamoDB table, covered in chapter 3, and the machine learning module, to be described briefly in chapter 10.) The full code is accessible from the book website in the **EXTRAS** tab.

We use the queue service's message attribute system to pass the article's title, abstract, and source in each message. (The arXiv data's source field provides enough information to train the predictor, but we do not use it for that purpose; instead, we append it to the prediction so that the predictor's accuracy can be evaluated by looking at the data stored in the table.)

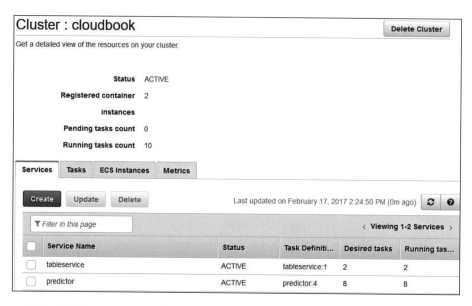

Figure 7.13: Eight instances of the predictor and two instances of the table service running.

The code to send the messages to the queue is also simple. We provide the complete code in notebook 10. We first load the data into three arrays and then send 100 messages using the Amazon SQS method **send_message**, as follows.

```
queue = sqs.get_queue_by_name(QueueName='bookque')
abstracts, sites, titles = load_docs("path-to-documents",
                                     "sciml_data_arxiv")
for i in range(1330,1430):
    queue.send_message(MessageBody='boto3', MessageAttributes ={
            'Title':{ 'StringValue': titles[i],
                      'DataType': 'String'},
            'Source':{ 'StringValue': sites[i],
                      'DataType': 'String'},
            'Abstract':{ 'StringValue': abstracts[i],
                      'DataType': 'String'}
        })
```

Table 7.1 shows the results of this action when using eight instances of the predictor microservice. As a further refinement, we could, as shown in figure 7.15 on page 120, partition the work performed by our microservice across two microservices: one to pull a task from queue and analyze it, and one to store a result. We can then mix and match different input queues and output services: for example, pulling from the Amazon queue, analyzing on Jetstream, and storing in Google Bigtable. We provide a version with this form on the book website in the **EXTRAS** tab.

```
import boto3, time
from socket import gethostname
from predictor import predictor
hostnm = gethostname()

# Create an instance of the ML predictor code
pred = predictor()

# Create instance of Amazon DynamoDB table (see chapter 3)
sqs = boto3.resource('sqs', region_name='us-west-2')
queue = sqs.get_queue_by_name(QueueName='bookque')

i = 0
while True:
    for message in queue.receive_messages(
            MessageAttributeNames=['Title', 'Abstract','Source']):
        timestamp = time.time()
        if message.message_attributes is not None:
            title = message.message_attributes.
                        get('Title').get('StringValue')
            abstract = message.message_attributes.
                        get('Abstract').get('StringValue')
            source = message.message_attributes.
                        get('Source').get('StringValue')
            predicted = pred.predict(abstract, source)

            metadata_item =
                {'PartitionKey': hostnm, 'RowKey': str(i),
                 'date' : str(timestamp), 'answer': source,
                 'predicted':  str(predicted), 'title': title}
            table.put_item(Item=metadata_item)
            message.delete()
            i = i+1
```

Figure 7.14: Abbreviated code for the prediction microservice.

Table 7.1: View of the DynamoDB table after processing 100 messages. The *Answer* column references the arXiv source, which can be used to verify the prediction.

PartitionKey	RowKey	Answer	Date	Predicted	Title
e0bfabe3d880	0	gr-qc	148...	Physics	Superconducting dark eng ...
e0bfabe3d880	1	physics.optics	148...	Physics	Directional out-coupling of il ...
e0bfabe3d880	2	q-bio.PI	148...	Bio	A guide through a family of p ...
e0bfabe3d880	4	math.PR	148...	Math	Critical population and error ...
e0bfabe3d880	5	physics.comp	148...	Phys	Coupling all-atom molecular ...
e0bfabe3d880	7	hep-th	148...	Pyysics	Nonsingular Cosmology from ...

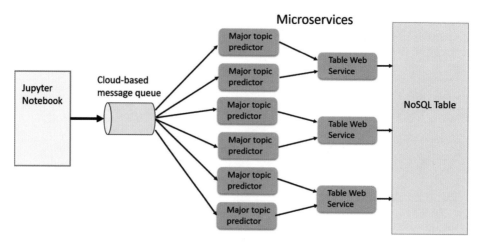

Figure 7.15: Version 2 of the document classifier in figure 7.11 partitions the microservice into a major topic selector service and a web service that handles the table storage.

Because putting a record into the table is much cheaper than running the data analysis predictor, we can increase throughput by replicating the predictor component. For example, in our Amazon implementation we ran on a two-server cluster, with one table service per server and multiple predictors per server. We have configured the table service to wait on a fixed TCP/IP port; hence there can be no more than one table service per server. Each predictor service simply posts its result to its local host at that port, providing a simple form of service discovery. With 20 predictors and two table services, the system can classify documents as fast as we can add them to the queue. (To scale further, we might want multiple table services per server, requiring a more sophisticated service discovery method.)

7.6.5   Google's Kubernetes

Google has run its major services using the microservice design for years. Recently, Google released a version of its underlying resource manager as open source, under the name **Kubernetes**. This service can be both installed on a third-party cloud and accessed within the Google Cloud. Creating a Kubernetes cluster on the Google Cloud is easy. Select the "Container Engine." On the "container clusters" page, there is a link that allows you to create a cluster. (With the free account you cannot make a large cluster: you are limited to about four dual-core servers.) Fill in the form and submit it, and you soon have a new cluster. Clicking on the icon >_ in the blue banner at the top of the form creates an instance of a "Cloud Shell" that is automatically authenticated to your account. Next, you must authenticate

your cloud shell with your new cluster. Select your container and click on the
connect button to the right to obtain the code to paste into the cloud shell. The
result should now look like figure 7.16.

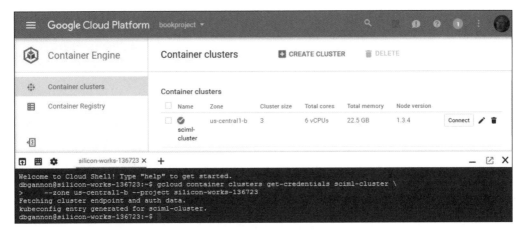

Figure 7.16: Kubernetes console and cloud shell connected to a small cluster.

You interact with Kubernetes, which is now running on our small cluster,
via command line calls entered into the cloud shell. Kubernetes has a different
and somewhat more interesting architecture than other container management
tools. The basic unit of scheduling in Kubernetes is the *pod*, a set of one or more
Docker-style containers together with a set of resources that are shared by the
containers in that pod. When launched, a pod resides on a single server or VM.
This approach has several advantages for the containers in that pod. Because the
containers in a pod all run on the same VM, they all share the same IP and port
space and thus find each other through conventional means such as *localhost*. They
can also share storage volumes that are local to the pod.

To start, let's run the notebook in a simple single-container pod. Use the
Kubernetes control command `kubectl` to run the following statements.

```
# Launch Jupyter and expose its port 8888
kubectl run jupyter --image=jupyter/scipy-notebook --port=8888
# To make visible externally, attach a load balancer
kubectl expose deployment jupyter --type=LoadBalancer
# Get service description
kubectl describe services jupyter
```

You can then get the IP address for Jupyter from the "LoadBalancer Ingress:"
field of the service description produced by the third command. (If that address is
not immediately visible, rerun the command.)

To duplicate the example in the preceding section, we need a queue service and a way to talk to it. Like Amazon, Google has an excellent message queue service, **Google Cloud Pub/Sub**, that supports both *push* and *pull* subscribers. However, rather than use this service here, we choose to demonstrate how to distribute the computation over the Internet by putting the queue service on a different cloud. Specifically, we deploy an instance of the open source queue service **RabbitMQ** `rabbitmq.com` on a VM running on Jetstream. We then use a Python package called **Celery** to communicate with the queue service.

Celery is a distributed remote procedure call system for Python programs. The Celery view of the world is that you have a set of worker processes running on remote machines and a client process that invokes functions that are executed on the remote machines. The workers and clients coordinate through a message broker running on Jetstream. Because Celery is a remote procedure call system, we define its behavior by creating functions that are annotated with a Celery object task.

When creating a Celery object, we must provide a name and a reference to the protocol being used, which in this case is the **Advanced Message Queuing Protocol** (AMQP). We can then run our predictor as follows; the command line argument provides a link to the specific RabbitMQ server.

```
>celery worker -A predictor -b 'amqp://guest@brokerIPaddr'
```

The code for our solution, shown in figure 7.17 on the following page, includes the needed components for interacting with the Google Cloud Datastore service that we described in chapter 3. The resources needed to run this example can be found on the book website in the **Extras** tab.

We can create as many instances of this microservice as we wish. If we create multiple instances, they share the load of handling the prediction requests. To invoke the remote procedure call from a client program, we use an `apply_async` function call. This involves creating a *stub* version of the function to define its parameters. For example, the following is a single invocation of our predictor.

```
from celery import Celery
app = Celery('predictor', broker='amqp://guest@brokerIPaddr',\
             backend='amqp')
@app.task
def predict(statement):
    return ["stub call"]
res = predict.apply_async(["this is a science document ..."])
print(res.get())
```

We have included here a reference to the RabbitMQ broker so that we can run this in a Jupyter notebook. The `apply_async` call returns immediately and

```
from celery import Celery
from socket import gethostname
from predictor import predictor
from gcloud import datastore
clientds = datastore.Client()
key = clientds.key('booktable')
hostnm = gethostname()
import time
app = Celery('predictor',broker='amqp://guest@brokerIPaddr', \
            backend='amqp')
pred = predictor()

# Define the functions that we will call remotely
@app.task
def predict(abstract, title, source):
    prediction = pred.predict(statement, source)
    entity = datastore.Entity(key=key)
    entity['partition'] = hostname'
    entity['date'] = str(time.time())
    entity['title' ] = title
    entity['prediction'] = str(prediction)
    clientds.put(entity)
    return [prediction]
```

Figure 7.17: An implementation of the document classifier that uses Kubernetes.

returns a *future* object. To get the actual returned value we have to wait, and we can do that with a call to `get()`. If we want to send a request to predict the classification of thousands of documents, we can do this as follows.

```
res = []
for doc in documents:
    res.append(predict.apply_async([doc])

# Now wait for them all to be done
predictions = [result.get() for result in res]
```

This statement dispatches a remote procedure call to the message broker. All our workers then participate in satisfying the requests. The result returned is a list of future objects. We can resolve them as they arrive, as shown.

Next we must ask Kubernetes to create and manage a collection of workers. First we must package the worker as a Docker container *cloudbook/predictor* and push it to the Docker Hub. As we did with ECS, the Amazon container service, we must create a task description, such as the following.

123

```
apiVersion: batch/v1
kind: Job
metadata:
    name:predict-job
spec:
    parallelism: 6
    template:
        metadata:
        name: job-wq
    spec:
        containers:
            - name: c
            image: cloudbook/predictor
            args: ["amqp://guest@brokerIPaddr"]
        restartPolicy: OnFailure
```

We package this document in a file `predict-job.json`. Notice that this task description contains a mix of the elements of the Amazon ECS task descriptor and the Amazon ECS service arguments. We can now ask Kubernetes to launch the worker set with the following command. Once they are up and running, they are ready to respond to requests. Kubernetes handles restarts in case of failures.

```
kubectl create -f predict-job.json
```

7.6.6 Mesos and Mesosphere

Mesosphere (from Mesosphere.com) is a data center operating system (DCOS) based on the original Berkeley Mesos system for managing clusters. We describe here how to install and use Mesosphere on Microsoft's Azure cloud. Mesosphere has four major components:

1. The **Apache Mesos** distributed system kernel.

2. The **Marathon** init system, which monitors applications and services and, like Amazon ECS and Kubernetes, automatically heals any failures.

3. **Mesos-DNS**, a service discovery utility.

4. **ZooKeeper**, a high-performance coordination service to manage the installed DCOS services.

When Mesosphere is deployed, it has a master node, a backup master, and a set of workers that run the service containers. Azure supports the Mesosphere components listed above as well as another container management service called Docker Swarm. Azure also provides a set of DCOS command line tools. For example, to see the full set of applications that Marathon is managing, we can issue the command `dcos marathon app list`, with the result shown in figure 7.18.

```
> dcos marathon app list
  ID               MEM    CPUS   TASKS   HEALTH   DEPLOYMENT   CONTAINER   CMD
  /nginx            16    0.1    1/1     ---      ---          DOCKER      None
  /rabbitsend3     512    0.1    0/0     ---      ---          DOCKER      None
  /spark          1024    1      1/1     1/1      ---          DOCKER      /sbin/init.sh
  /storm-default  1024    1      2/2     2/2      ---          DOCKER      ./bin/run-
                                                                          with-marathon.sh
  /zeppelin       2048    1      1/1     1/1      ---          DOCKER      sed ...
```

Figure 7.18: DCOS command line example.

Mesosphere also provides excellent interactive service management consoles. When you bring up Mesos on Azure through the Azure Container Services, the console presents a view of your service health, current CPU and memory allocations, and current failure rate. If you next select the services view and then the Marathon console, you see the details of the applications that Marathon is managing, as shown in figure 7.19. You see that from a previous session that we currently have running one instance of NGINX, one instance of Spark, two instances of Storm, and one instance of Zeppelin.

Figure 7.19: View of the Marathon console.

The process of launching applications composed of Docker containers is similar to that used in Amazon ECS and Kubernetes. We start with a task description

JSON file such as the following, which specifies the type of container, the Dockerfile to use, the number of instances required, and other information.

```
{
    "container": {
        "type": "DOCKER",
        "docker": {
            "image": "cloudbook/predictor"
        }
    },
    "id": "worker",
    "instances": 1,
    "cpus": 0.2,
    "mem": 512,
}
```

Next, send this descriptor to Marathon with the following command.

```
dcos marathon app add config.json
```

Notice that this task descriptor specifies only one instance. You can issue the following command to increase this instance count to nine.

```
dcos marathon app update worker env='{"instances":"9"}'
```

We now return to our document classifier example. We completed a version of the full classifier based on the concepts in figure 7.10. The actual implementation splits the predictor service into two parts: the core document classsifier and a service that pushes the data into an Azure table. Once again, we use RabbitMQ as a message broker, as shown in figure 7.20 on the following page.

In this case the first stage of classification performs the initial partition into the main subject headings; depending on the results, it then pushes the document into a queue specific to the main subject. We assigned one or more instances of the classifier to each topic queue, with numbers determined by the size of that main topic queue. (For example, physics has more documents than any other subject, and thus we assigned that topic more instances.) The subarea classifiers are based on the same code base as the main topic classifiers. They specialize their behavior by loading trained data models specific to that subject and its subareas. A detailed analysis of system performance as a function of the number of classifier instances is available online [133]. In brief, using nine servers and computing speedup relative to the time for a single classifier to compute the analysis, the system achieved a speedup of about 8.5 using 16 classifier instances.

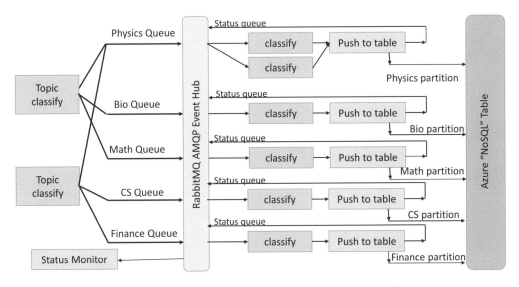

Figure 7.20: Diagram of the full subject area classifier.

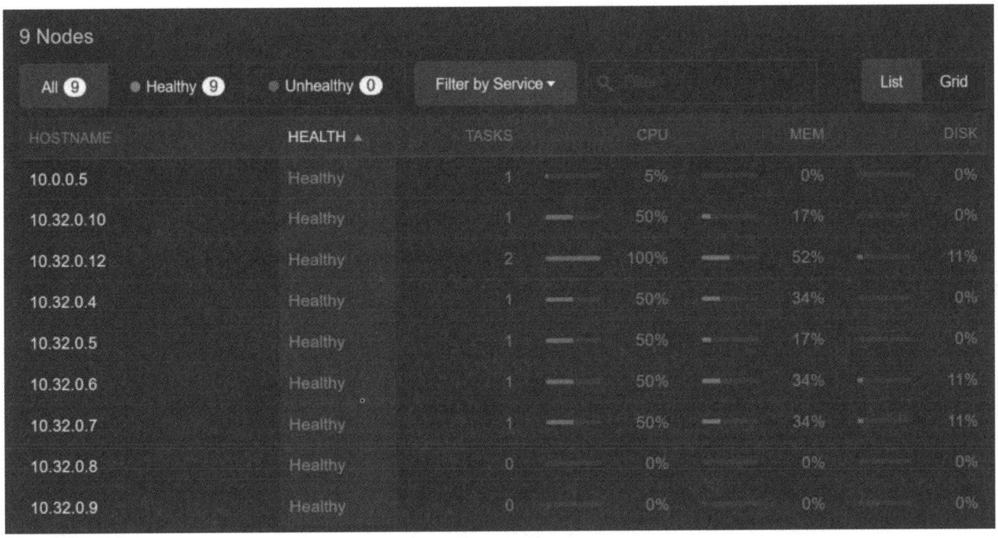

Figure 7.21: DCOS system status showing the load on each server in the cluster.

The Azure DCOS container service status tool provides a useful view of system status. Figure 7.21 shows the load on each server when eight classifier services are running. Notice that one server has two instances while two servers are empty. This apparent imbalance is a feature, not a bug: at one point during the run, two services crashed, causing DCOS to migrate the failed instances to other nodes.

7.7 HTCondor

The **HTCondor** `research.cs.wisc.edu/htcondor` high-throughput computing system is a particularly mature technology for scientific computing in the cloud. The following are examples of HTCondor applications.

The **Globus Genomics** system [187] uses HTCondor to schedule large numbers of bioinformatics pipelines on the Amazon cloud. We describe this system in more detail in section 14.4 on page 303, as an example of how to build scalable SaaS. **GeoDeepDive** `geodeepdive.org`, part of the NSF EarthCube project, is an infrastructure for text and data mining that uses HTCondor for large analyses of massive text collections. **Pegasus** [109] is a workflow system for managing large scientific computations on top of HTCondor. A collaboration between Google and Fermilab's **HEPCloud project** used HTCondor to process data from a high energy physics experiment. The Google Cloud *Preemptible Virtual Machines* provided 160,000 virtual cores and 320 TB of memory at a cost of only $1,400 per hour. With the data stored in Google Cloud Storage, the tasks spawned by HTCondor on the VMs read data using **gcfuse**, a mechanism to mount storage data buckets as Linux file systems. The output was returned to Fermilab over the U.S. Department of Energy's ESnet network.

7.8 Summary

The development of the digital computer led computation to join experiment and theory as a third major paradigm of scientific discovery and engineering design. Supercomputers became essential lab instruments. In the 40 years that followed, the power of supercomputers has grown by a factor of over one billion. But now science is rapidly evolving into a **fourth paradigm** [153] of data-driven discovery.

The technology companies whose business depends on understanding their online customers are the same companies that have built the cloud to handle data analytics workloads. As a result, data science and machine learning are now among the most in-demand technical specialties [215]. To meet the needs of "big data" analysis, cloud builders are rapidly evolving cloud data centers architectures down paths pioneered by the supercomputer vendors.

Taking advantage of large-scale parallelism was a first step to using the cloud for analytics. Bulk synchronous data parallelism via Hadoop and graph-driven execution are standard approaches to extracting enough concurrency to pull knowledge from mountains of data. Running and managing massive online systems also requires new approaches to parallelism that have made it easier to deploy and maintain thousands of processes. Containers and microservice architectures have become the basis for new cloud software engineering paradigms.

As public clouds evolved, their customers demanded better and simpler ways of using the cloud to compute at ever larger scales. In response, public cloud vendors introduced the services discussed in this chapter. Some customers need traditional HPC capabilities to run legacy HPC MPI codes. As we have illustrated, Amazon and Microsoft now have services that customers can use to build medium-scale HPC clusters on demand. While still far from the performance level of the latest supercomputers, these cloud solutions can be deployed quickly, used for a specific task, and then shut down, all for a tiny fraction of the cost.

Other cloud customers need less tightly coupled computing. Amazon, Azure, and Google have excellent scale-out container management services that are easy to use and manage. As the importance of machine learning has grown in the tech industry, so too has the need to train massive, deep neural networks. This need has driven the growth of GPU-based computing in the cloud. First used internally by some cloud vendors for deep learning, this technology has now emerged as a new class of multiGPU server instances available to anybody. Combining multiGPU servers with the ability to build a custom HPC cluster makes it possible to deploy truly powerful on-demand computing platforms.

In the chapters that follow we return to the theme of scale as it relates to data analysis, machine learning, and event processing.

7.9 Resources

We recommend the excellent tutorial by Sebastien Goasguen [140] on how to deploy and manage Kubernetes from Python. Microsoft recently released the Azure Batch Shipyard [6] open source tools for managing clusters of containers.

In addition to notebook 9 and notebook 10, introduced in the text, the source code for the C program in the Amazon CfnCluster example from section 7.2.3 on page 100 is on the book website in the **Extras** tab, as are the source code and data for the Amazon science document classification example, and the source code for the simple Docker example of section 7.6.5 on page 120.

For those interested in the evolution of cloud data center networks, we recommend Greenberg et al.'s review of the challenges inherent in building networks for cloud data centers [145] and two fascinating papers by Google engineers describing the techniques used to organize Google's data center networks [160, 236].

Part III

The Cloud as Platform

Building your own cloud
What you need to know
Using Eucalyptus
Part IV   Using OpenStack

Security and other topics
Securing services and data
Solutions
History, critiques, futures   **Part V**

Part III   **The cloud as platform**

Data analytics	Streaming data	Machine learning	Research data portals
Spark & Hadoop	Kafka, Spark, Beam	Scikit-Learn, CNTK,	DMZs and DTNs, Globus
Public cloud Tools	Kinesis, Azure Events	Tensorflow, AWS ML	Science gateways

Managing data in the cloud
File systems
Object stores
Databases (SQL)
NoSQL and graphs
Warehouses
Part I   Globus file services

Computing in the cloud
Virtual machines
Containers – Docker
MapReduce – Yarn and Spark
HPC clusters in the cloud
Mesos, Swarm, Kubernetes
Part II   HTCondor

Part III:
The Cloud as Platform

"May your mountains rise into and above the clouds."
—Edward Abbey

As we noted in chapter 1, the cloud is a lot more than a virtual computer: it is a rich ecosystem of services that can slash the expertise, time, and money required to build sophisticated applications. For example, say that you want to build an application to monitor environmental sensors and alert you when a certain combination of sensors indicates anomalous behavior. Or perhaps you need to explore a large archive of image data for samples to train a deep neural network, so that you can analyze a much larger archive. Or you have been charged with building a system to deliver large quantities of genomic data to collaborators around the world. Each of these tasks sounds like a massive task, but as we will see, cloud services can make each of them surprisingly easy.

In this regard, a cloud serves as a platform: an environment that allows you to develop, run, and manage applications without the need to set up, run, and maintain the hardware and software infrastructure that would otherwise be needed to host those applications. In the environmental sensor example, you can receive, enqueue, and process events without having to write the complex software that would normally be used to perform those tasks. Your application can, furthermore, scale to handle more events automatically, without you having to implement specialized load balancing logic. Need access control? Archiving? Auditing? Data analytics? Each of these capabilities is easily available.

This concept of a platform is not new to science and engineering. Many people use Matlab, Mathematica, SPSS, SAS, R, or Python, each of which provides capabilities that simplify the development of certain classes of application. When hosted on the cloud, these same tools can become collaborative information-processing laboratories.

In general, then, a cloud platform comprises a set of software components that are operated by the cloud provider and that software developers can incorporate into their applications, for example by REST API calls. Many systems satisfy this broad definition, and a surprising number of those systems have been used in science and engineering in one way or another. (Scientists and engineers are enterprising people!) For example, Facebook provides a set of programming interfaces and tools that developers can use to integrate with the "social graph" that Facebook maintains of personal relations and information. Researchers have used this platform's capabilities to implement scientific collaboration systems and even peer-to-peer resource-sharing systems in which Facebook friends share storage space on their computers [88]. The Twitter and Salesforce platforms have seen similar use.

The number of cloud platform capabilities is so large that we cannot hope to do them justice here. Instead, we focus on four classes of cloud platform services:

- **Data analytics**, as implemented with the Hadoop and YARN tools including Spark. We show how data analytics can be used on Amazon Elastic MapReduce and Azure HDInsight and Google's Cloud Datalab. We also look at data warehouse tools such as Azure Data Lake and Amazon Athena.

- **Streaming data** services, which have become a fully integrated part of the public cloud landscape. Amazon Kinesis and its analytics tools, along with Azure Event Hubs and Stream Analytics, are easily used and powerful. The open source community also has developed a rich collection of tools for monitoring and analyzing streaming data.

- **Machine learning** services, which combine open source libraries and interactive cloud-based development environments to provide exciting new capabilities. Deep learning is revolutionizing the field because of the availability of extremely large data collections and powerful computing platforms.

- **Globus platform services**, which provide identity, group, and research data management capabilities that simplify the development of applications and systems that integrate people and data at disparate locations, such as research data management portals.

Chapter 8

Data Analytics in the Cloud

"Science is what we understand well enough to explain to a computer.
Art is everything else we do."

—Donald Knuth

What we know today as public clouds were originally created as internal data centers to support services such as e-commerce, e-mail, and web search, each of which involved the acquisition of large data collections. To optimize these services, companies started performing massive amounts of analysis on those data. After these data centers morphed into public clouds, the methods developed for these analysis tasks were increasingly made available as services and open source software. Universities and other companies have also contributed to a growing collection of excellent open source tools. Together this collection represents an enormous ecosystem of software supported by a large community of contributors.

The topic of data analytics in the cloud is huge and rapidly evolving, and could easily fill an entire volume itself. Furthermore, as we observed in chapter 7, science and engineering are themselves rapidly evolving towards a new data-driven fourth paradigm of data-intensive discovery and design [153]. We survey some of the most significant approaches to cloud data analytics and, as we have done throughout this book, leave you with examples of experiments that you can try on your own.

We begin with the first major cloud data analysis tool, Hadoop, and describe its evolution to incorporate Apache YARN. Rather than describe the traditional Hadoop MapReduce tool, we focus on the more modern and flexible Spark system. Both Amazon and Microsoft have integrated versions of YARN into their standard service offerings. We describe how to use Spark with the Amazon version of YARN,

Amazon Elastic MapReduce, which we use to analyze Wikipedia data. We also present an example of the use of the Azure version of YARN, HDInsight.

We then turn to the topic of analytics on truly massive data collections. We introduce Azure Data Lake and illustrate the use of both Azure Data Lake Analytics and Amazon's similar Athena analytics platform. Finally, we describe a tool from Google called Cloud Datalab, which we use to explore National Oceanographic and Atmospheric Administration (NOAA) data.

8.1   Hadoop and YARN

We have already introduced Hadoop and the MapReduce concept in chapter 7. Now is the time to go beyond an introduction. When Hadoop was introduced, it was widely seen as the tool to use to solve many large data analysis problems. It was not necessarily efficient, but it worked well for extremely large data collections distributed over large clusters of servers.

The **Hadoop Distributed File System** (HDFS) is a key Hadoop building block. Written in Java, HDFS is completely portable and based on standard network TCP sockets. When deployed, it has a single **NameNode**, used to track data location, and a cluster of **DataNodes**, used to hold the distributed data structures. Individual files are broken into 64 MB blocks, which are distributed across the DataNodes and also replicated to make the system more fault tolerant. As illustrated in figure 8.1 on the following page, the NameNode keeps track of the location of each file block and the replicas.

HDFS is not a POSIX file system: it is write-once, read-many, and only eventually consistent. However, command line tools make it usable in a manner similar to a standard Unix file system. For example, the following commands create a "directory" in HDFS, pull a copy of Wikipedia from a website, push those data to HDFS (where they are blocked, replicated and stored), and list the directory.

```
$hadoop fs -mkdir /user/wiki
$curl -s -L http://dumps.wikimedia.org/enwiki/...multisream.xml.bz2\
   | bzip2 -cd |hadoop fs -put - /user/wiki/wikidump-en.xml
$hadoop fs -ls /user/wiki
  Found 1 items
  -rw-r--r-- hadoop 59189269956 21:29 /user/wiki/wikidump-en.xml
```

Hadoop and HDFS were originally created to support only Hadoop MapReduce tasks. However, the ecosystem rapidly grew to include other tools. In addition, the original Hadoop MapReduce tool could not support important application classes

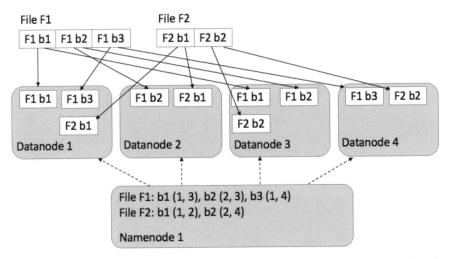

Figure 8.1: Hadoop Distributed File System with four DataNodes and two files broken into blocks and distributed. The NameNode keeps track of the blocks and replicas.

such as those requiring an iterative application of MapReduce [81] or the reuse of distributed data structures.

Apache **YARN** (Yet Another Resource Negotiator) represents the evolution of the Hadoop ecosystem into a full distributed job management system. It has a resource manager and scheduler that communicates with node manager processes in each worker node. Applications connect to the resource manager, which then spins up an application manager for that application instance. As illustrated in figure 8.2 on the next page, the application manager interacts with the resource manager to obtain "containers" for its worker nodes on the cluster of servers. This model allows multiple applications to run on the system concurrently.

YARN is similar in many respects to the Mesos system described in chapter 7. The primary difference is that YARN is designed to schedule MapReduce-style jobs, whereas Mesos is designed to support a more general class of computations, including containers and microservices. Both systems are widely used.

8.2 Spark

Spark's design addresses limitations in the original Hadoop MapReduce computing paradigm. In Hadoop's linear dataflow structure, programs read input data from disk, map a function across the data, reduce the results of the map, and store reduction results on disk. Spark supports a more general graph execution model

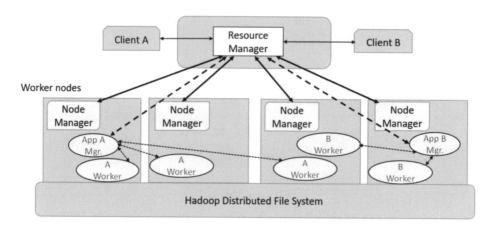

Figure 8.2: YARN distributed resource manager architecture.

that allows for iterative MapReduce as well as more efficient data reuse. Spark is also interactive and much faster than pure Hadoop. It runs on both YARN and Mesos, as well as on a laptop and in a Docker container. In the following paragraphs we provide a gentle introduction to Spark and present some data analytics examples that involve its use.

A central Spark construct is the **Resilient Distributed Dataset (RDD)**, a data collection that is distributed across servers and mapped to disk or memory, providing a restricted form of distributed shared memory. Spark is implemented in **Scala**, an interpreted, statically typed object-functional language. Spark has a library of Scala parallel operators, similar to the Map and Reduce operations used in Hadoop, that perform transformations on RDDs. (The library also has a nice Python binding.) More precisely, Spark has two types of operations: **transformations** that map RDDs into new RDDs and **actions** that return values to the main program: usually the read-eval-print-loop, such as Jupyter.

8.2.1   A Simple Spark Program

We introduce the use of Spark by using it to implement a trivial program that computes an approximation to π via this identity:

$$\lim_{n->\infty} \sum_{i=1}^{n} \frac{1}{i^2} = \frac{\pi^2}{6} \tag{8.1}$$

Our program, shown in figure 8.3 on the following page and in notebook 11, uses a map operation to compute $\frac{1}{i^2}$ for each of n values of i, and a reduce operation to sum the results of those computations. The program creates a one-dimensional array of integers that we then convert to an RDD partitioned into two pieces. (The array is not big, but we ran this example on a dual-core Mac Mini.)

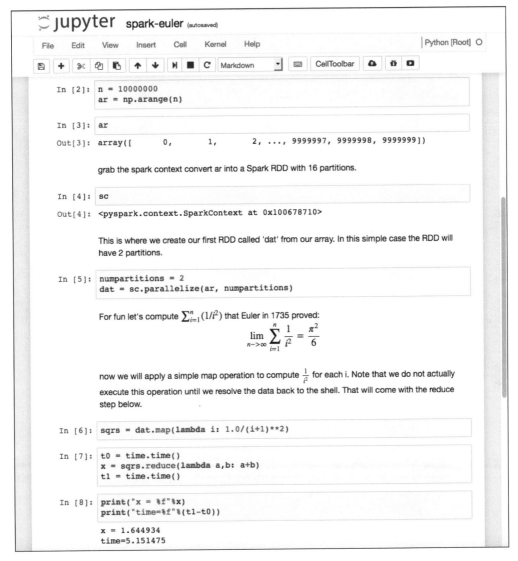

Figure 8.3: Computing with Spark.

In Spark, the partitions are distributed to the workers. Parallelism is achieved by applying the computational parts of Spark operators on each partition in parallel, using, furthermore, multiple threads per worker. For actions, such as a reduce, most of the work is done on each partition and then across partitions as needed. The Spark Python library exploits Python's ability to create anonymous functions using the lambda operator. Code is generated for these functions, and they can then be shipped by Spark's work scheduler to the workers for execution on each RDD partition. In this case we use a simple MapReduce computation.

8.2.2 A More Interesting Spark Program: K-means Clustering

We now consider the more interesting example of k-means clustering [150]. Suppose you have 10,000 points on a plane, and you want to find k new points that are the centroids of k clusters that partition the set. In other words, each point is to be assigned to the set corresponding to the centroid to which it is closest. Our Spark solution is in notebook 12.

We use an array `kPoints` to hold the k centroids. We initialize this array with random values and then apply an iterative MapReduce algorithm, repeating these two steps until the centroids have not moved far from their previous position:

1. For each point, find the index of the centroid to which it is nearest and assign the point to the cluster associated with that nearest centroid.

2. For each cluster, compute the centroid of all points in that cluster, and replace the previous centroid in `kPoints` with that new centroid.

We first define the following function to compute, for a given point p, the index of the centroid in `kPoints` to which p is nearest.

```
def closestPoint(p, kPoints):
    bestIndex = 0
    closest = float("+inf")
    for i in range(len(kPoints)):
        tempDist = np.sum((p - kPoints[i]) ** 2)
        if tempDist < closest:
            closest = tempDist
            bestIndex = i
    return bestIndex
```

The locations of our 10,000 points are in an array called `data`. We use the following map expression to create, as a new RDD, the set of tuples $(j, (p, 1))$ in which j is the index of the closest centroid for each p in `data`.

```
data.map( lambda p: (closestPoint(p, kPoints), (p, 1)))
```

This use of tuples of form $(p, 1)$ is a common MapReduce idiom. We want to compute for each j the sum of all tuples $(p, 1)$, to obtain tuples of the form

$$(j, (\sum p, \sum 1)).$$

To do this, we use a `reduceByKey` operation as follows.

```
reduceByKey(lambda x, y : (x[0] + y[0], x[1] + y[1]))
```

The sum of the 1s is just the count of the j tuples in our set, so we can compute the centroid by dividing the sum of the ps by this count. This is an RDD of size k, and we can collect it and use it as the `kPoints` for the next iteration. The full code is now in an iteration as follows.

```
tempDist = 1.0
while tempDist > convergeDist:
    newPoints = data \
            .map( lambda p: (closestPoint(p, kPoints), (p, 1))) \
            .reduceByKey(lambda x, y : (x[0] + y[0], x[1] + y[1])) \
            .map(lambda x : (x[0], x[1][0]/ x[1][1])) \
            .collect()

    tempDist = sum(np.sum((kPoints[i] - y) ** 2)   \
                    for (i, y) in newPoints)
    for (i, y) in newPoints:
        kPoints[i] = y
```

To summarize: We have a map, followed by a reduce-by-key, followed by another map, and finally a collect that brings the values of newPoints back to the read eval-print-loop. Each Spark operation is executed on the cluster of cores in which the RDD is located. In fact, what the Python program is doing is compiling a graph that is then executed by the Spark engine.

We note that this example was designed to illustrate some of the standard Spark idioms, and it is not the best k-means algorithm. Spark has a machine learning library with a much better implementation.

8.2.3   Spark in a Container

Running a containerized version of Spark on your laptop is straightforward. You can also easily run Spark on a remote VM with many cores: as long as the VM has Docker installed, you can follow the same procedure described in section 6.2 on page 87 but with a different version of Jupyter.

```
docker run -e GEN_CERT=yes -d -p 8888:8888 \
        -v /tmp/docmnt:/home/jovyan/work/docmnt \
      jupyter/all-spark-notebook start-notebook.sh \
      --NotebookApp.password='sha1:....'
```

When developing the k-means example, provided as notebook 12, we used a container with a host drive **/vol1/dockerdata** with 10 GB of memory and cores 0, 1, 2, and 3. We created this container as follows.

```
docker run -e GEN_CERT=yes -d -p 8888:8888 --cpuset-cpus 0-3 -m 10G\
        -v /tmp/docmnt:/home/jovyan/work/docmnt \
      jupyter/all-spark-notebook start-notebook.sh \
      --NotebookApp.password='sha1:....'
```

8.2.4 SQL in Spark

Python and Spark can also execute SQL commands [64]. We illustrate this capability in notebook 13 with a simple example. We have a comma-separated value (CSV) file, **hvac.csv**, with a header and three data lines, as follows.

```
Date, Time, desired temp, actual temp, buildingID
3/23/2016, 11:45, 67, 54, headquarters
3/23/2016, 11:51, 67, 77, lab1
3/23/2016, 11:20, 67, 33, coldroom
```

We load this file into Spark and create an RDD by applying the **textFile** operator to the Spark context object. We convert the text file RDD into an RDD of tuples by stripping off the header and mapping the rest to typed tuples. We create an SQL context object and schema type, and then an SQL DataFrame [45] (see section 10.1 on page 192), **hvacDF**.

```
from pyspark.sql.types import *
hvacText = sc.textFile("/pathto/file/hvac.csv")
hvac = hvacText.map(lambda s: s.split(",")) \
            .filter(lambda s: s[0] != "Date") \
            .map(lambda s:(str(s[0]), str(s[1]),
                        int(s[2]), int(s[3]), str(s[4]) ))
sqlCtx = SQLContext(sc)
hvacSchema = StructType([StructField("date", StringType(), False),
                    StructField("time", StringType(), False),
                    StructField("targettemp", IntegerType(), False),
                    StructField("actualtemp", IntegerType(), False),
                    StructField("buildingID", StringType(), False)])
hvacDF = sqlCtx.createDataFrame(hvac, hvacSchema)
```

We are now ready to execute SQL operations on our data. For example, we can use the `sql()` method to extract a new SQL RDD DataFrame consisting of the `buildingID` column with the following command.

```
x = sqlCtx.sql('SELECT buildingID from hvac')
```

Yet more fun is to use Jupyter and IPython **magic operators**, which allow you to define small extensions to the language. Using an operator developed by Luca Canali, we can create a magic operator that allows us to enter the SQL command in a more natural way and have the results printed as a table. Using this new magic operator `%%sql_show`, we can create a table consisting of the buildingID, the data, and the difference between the desired and actual temperatures.

```
%%sql_show
SELECT buildingID ,
       (targettemp - actualtemp) AS temp_diff ,
       date FROM hvac
WHERE date = "3/23/2016"

+------------+---------+---------+
|  buildingID|temp_diff|     date|
+------------+---------+---------+
|headquarters|       13|3/23/2016|
|        lab1|      -10|3/23/2016|
|    coldroom|       34|3/23/2016|
+------------+---------+---------+
```

We provide details on the SQL magic operators and a link to Canali's blog in notebook 13.

8.3 Amazon Elastic MapReduce

Deploying Hadoop on a cluster used to require a team of systems professionals. Fortunately, products from Amazon, Microsoft, Google, IBM, and others have largely automated this task. Amazon's **Elastic MapReduce (EMR)** service makes creating a YARN cluster trivial. All you need to do is select your favorite combination of tools from their preconfigured lists, specify the instance type and number of worker nodes that you want, set up your usual security rules, and click `Create cluster`. In about two minutes you are up and running.

The configuration that we found most attractive combines Spark, YARN, and an interactive web-based notebook tool called **Zeppelin** `zeppelin.apache.org`. Zeppelin is similar to Jupyter, has an impressive user interface and graphics capabilities, and is highly integrated with Spark. For consistency, however, we stick with Jupyter in our presentation here.

When the EMR cluster comes up, it is already running one instance of Spark on YARN as well as HDFS. Getting Jupyter installed and running with YARN take a few extra commands, as shown in notebook 14.

To illustrate the use of Spark on EMR, we count how often famous people's names occur in Wikipedia. The program is in figure 8.4. We first load a small sample of Wikipedia access logs—from 2008 to 2010—from S3. Since this file contains only about four million records, our text-file RDD initially has just one partition. Thus, we next repartition the RDD into 10 segments, so that we can better exploit the parallelism in the Spark operators in subsequent steps. Each line of the text file comprises an identifier, the name of the page accessed, and an associated access count, separated by blank characters. To make these data easier to work with, we transform each line into an array by splitting on blank characters.

To count the number of hits for each person, we define a function `mapname` that returns the name of the person in the page title, and execute a map to replace each row with a new pair consisting of the name and the value in the count field for that row. We then reduce by the person name and add the hit counts. To look at the list ranked by hit count, we execute the pipeline defined by RDD `remapped` with a version of the `take()` function as follows.

```
remapped.takeOrdered(20, key = lambda x: -x[1])

[('Lady_Gaga', 4427),
 ('Bill_Clinton', 4221),
 ('Michael_Jackson', 3310),
 ('Barack_Obama', 2518),
 ('Justin_Bieber', 2234),
 ('Albert_Einstein', 1609),
 ('Byron', 964),
 ('Karl_Marx', 892),
 ('Arnold_Schwarzenegger', 820),
 ('Bill_Gates', 799),
 ('Steve_Jobs', 613),
 ('Vladimir_Putin', 563),
 ('Richard_Nixon', 509),
 ('Vladimir_Lenin', 283),
 ('Donald_Trump', 272),
 ('Nicolas_Sarkozy', 171),
 ('Hillary_Clinton', 162),
 ('Groucho_Marx', 152)]
 ('Werner_Heisenberg', 92),
 ('Elon_Musk', 21)]
```

```
# Define list of famous names
namelist = ['Albert_Einstein', 'Lady_Gaga', 'Barack_Obama',
    'Richard_Nixon','Steve_Jobs', 'Bill_Clinton', 'Bill_Gates',
    'Michael_Jackson','Justin_Bieber', 'Vladimir_Putin',
    'Byron', 'Donald_Trump', 'Hillary_Clinton', 'Nicolas_Sarkozy',
    'Werner_Heisenberg', 'Arnold_Schwarzenegger', 'Elon_Musk',
    'Vladimir_Lenin', 'Karl_Marx', 'Groucho_Marx']

# Transform a line into an array by splitting on blank characters
def parseline(line):
    return np.array([x for x in line.split(' ')])

# Filter out lines not containing famous name in the page title
def filter_fun(row, titles):
    for title in titles:
        if row[1].find(title) > -1:
            return True
    else:
        return False

# Return name of person in page title
def mapname(row, names):
    for name in names:
        if row[1].find(name) > -1:
            return name
    else:
        return 'huh?'

# ------ Load and process data ---------------------------------------
# Load Wikipedia data from S3
rawdata = sc.textFile( \
        "s3://support.elasticmapreduce/bigdatademo/sample/wiki")

# Repartition initial RDD into 10 segments, for parallelism
rawdata = rawdata.repartition(10)

# Split each line into an array
data = rawdata.map(parseline)

# Filter out lines without a famous name
filterd = data.filter(lambda p: filter_fun(p, namelist))

# Map: Replace each row with (name, count) pair.
# Reduce by name: Add counts
remapped =filterd.map(lambda row:(mapname(row,namelist),int(row[2])))
                .reduceByKey(lambda v1, v2: v1+v2)
```

Figure 8.4: Our program for counting famous people in Wikipedia.

We discover that pop stars are more popular than future presidential candidates, at least during 2008 to 2010. We choose not to pursue this research further.

To conclude this example, we load the full Wikipedia dump and make a list of all of the main pages. We have stored the full Wikipedia dump file in HDFS using the code presented in section 8.1 on page 136. The dump is a 64 GB file with each line containing one line from an XML file. We can load it directly from HDFS as follows using the `hdfs:///` prefix. The file has over 900 million lines. When loaded, it comprises 441 partitions in the RDD.

```
wikidump = sc.textFile("hdfs:///user/wiki/wikidump-en.xml")
wikidump.count()
927769981
wikidump.getNumPartitions()
441
```

To determine the titles of each Wikipedia entry, we filter for the lines containing XML tag `<title>`. We look at the first 12 of the more than 17 million entries.

```
def findtitle(line):
    if line.find('<title>') > -1:
        return True
    else:
        return False

titles = wikidump.filter(lambda p: findtitle(p))
titles.count()
17008269
titles.take(12)

[u'    <title>AccessibleComputing</title>',
 u'    <title>Anarchism</title>',
 u'    <title>AfghanistanHistory</title>',
 u'    <title>AfghanistanGeography</title>',
 u'    <title>AfghanistanPeople</title>',
 u'    <title>AfghanistanCommunications</title>',
 u'    <title>AfghanistanTransportations</title>',
 u'    <title>AfghanistanMilitary</title>',
 u'    <title>AfghanistanTransnationalIssues</title>',
 u'    <title>AssistiveTechnology</title>']
 u'    <title>AmoeboidTaxa</title>',
 u'    <title>Autism</title>',
```

To do more with these data, you need to assemble each of the 17 million XML records for each page into single items. An exercise for the ambitious reader!

8.4   Azure HDInsight and Data Lake

Microsoft has long had a MapReduce framework called Cosmos that is used as the main data analytics engine for many internal projects such as the Bing search engine. Cosmos is based on a directed graph execution model and is programmed in an SQL-like language called SCOPE. While Cosmos was a candidate for release as a general MapReduce tool for Azure users, the company decided that the Hadoop/YARN ecosystem was of such great interest to their customers that they would keep Cosmos internal and support YARN as the public offering.

Called **HDInsight** on Azure, this service supports Spark, Hive, HBase, Storm, Kafka, and Hadoop MapReduce, and is backed by a guarantee of 99.9% availability. Programming tools include integration into Visual Studio, Python, R, Java, and Scala as well as .Net languages. HDInsight is based on the Hortonworks Data Platform distribution integrated with Azure security. All of the usual Hadoop and YARN components are there, including HDFS as well as tools that integrate other Microsoft business analytics tools such as Excel and SQL Server.

Creating an HDInsight cluster with Spark and Jupyter already installed is easy. As with many Azure services, there is a preconfigured template that you can use. The complete instructions are online [194]. Clicking on the *Deploy to Azure* link in the documents takes you to your Azure account and sets up the script. You only need to fill in a few standard account details, such as a name for your cluster, the login id, and the password for the SSH and master nodes. After a few minutes the cluster is up. Going to the Azure portal, look for your new cluster and click on the name. You should see an image like that in figure 8.5 on the next page.

Clicking on the HDInsight Cluster Dashboard icon provides you with live status data about the cluster, such as memory, network, and CPU use. Clicking on the Jupyter icon first authenticates you and then takes you to the Jupyter home, where you have a choice of PySpark or Scala notebooks. Both directories contain excellent tutorial examples.

HDInsight uses a version of HDFS that is implemented on top of standard Azure Blob storage. You can load a file with a command like the following.

```
newRDD = spark.sparkContext.textFile('wasb:///mycontainer/file.txt')
```

Rather than go through another Spark example here, we defer further discussion to chapter 10, where we use Spark in an HDInsight machine learning example.

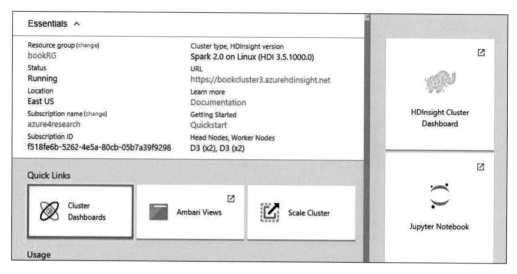

Figure 8.5: Azure portal view of a HDInsight cluster.

8.4.1 Azure Data Lake Store

HDInsight is part of something larger called **Azure Data Lake**, as shown in figure 8.6 on the following page. Data Lake also includes the Azure Data Lake Store, a data warehouse for petascale data collections. This store can be thought of as a giant extension of HDFS in the cloud. It does not have the 500 TB storage limits of Azure blobs, and is designed for massive throughput. It supports both structured and unstructured data. The access protocol for the Data Lake Store, **WebHDFS**, is the REST API for HDFS. Consequently you can access the Data Lake Store from anywhere, and with the same commands that you use to access HDFS. A Python SDK [7] allows access to the store from Jupyter or via a command line tool using familiar HDFS commands, as shown below.

```
> python azure/datalake/store/cli.py
azure> help

Documented commands (type help <topic>):
========================================
cat    chmod  close  du    get     help  ls    mv     quit  rmdir
touch  chgrp  chown  df    exists  head  info  mkdir  put   rm
tail
azure>
```

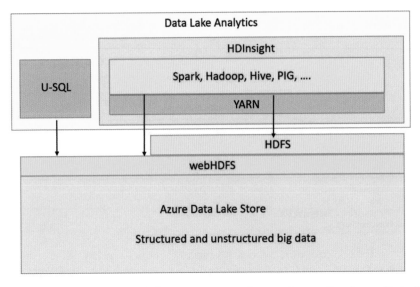

Figure 8.6: Conceptual view of the components that make up the Azure Data Lake.

8.4.2   Data Lake Analytics

Azure Data Lake Analytics consists of HDInsight and associated tools (Spark, Hadoop, etc.), plus a data analytics tool called **U-SQL** for scripting large analytics tasks. U-SQL combines SQL queries and declarative program functions written in C#, Python, or R. U-SQL is intended for massively parallel analysis of terabyte-to petabyte-size data collections. When you write a U-SQL script, you are actually building a graph. As shown in figure 8.7 on the next page, the program is a graph of queries in which regular declarative program functions can be embedded.

When you run a U-SQL job, you specify the degrees of parallelism that you wish to exploit for each of the query tasks in the computation. The result is a directed acyclic graph in which some nodes are parallelized according to your suggestions. You do not explicitly allocate VMs or containers, and you are charged for the total execution time of the graph nodes.

8.5   Amazon Athena Analytics

Athena, a recent addition to the Amazon analytics toolbox, is designed to allow users to query data in Amazon S3 without having to launch VM instances. Like Data Lake Analytics, Athena is an example of the concept of serverless computing that we introduced briefly in chapter 4.

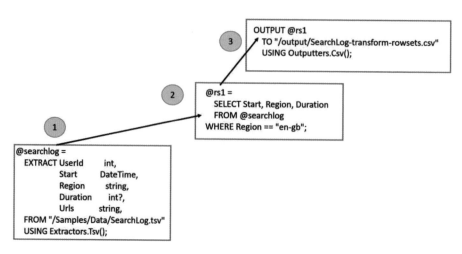

Figure 8.7: A U-SQL program defines a graph of queries. In the first step an object `searchlog` is extracted from a file. In step two the object `rs1` is created from `searchlog` by extracting items with region "en-gb". In the final step the object `rs1` is converted to a CSV file and saved.

Athena is accessed via the Amazon portal, which provides an interactive query tool based on standard SQL. You first put your data in S3 in one of several standard forms, including text files, CSV files, Apache web logs, JSON files, and a column-oriented format such as **Apache Parquet**. (Parquet files store the columns of a data table in contiguous locations, suitable for efficient compression and decompression into distributed file systems like HDFS. They are easily generated and consumed by any of the Hadoop ecosystem of tools, including Spark.)

The Athena query editor allows you to define the schema for your data and the method to be used to extract the data from the S3 source. Athena treats data in S3 as read-only, and all transformations are made in the internal Athena engine and storage. Once you have your data defined, you can explore them with the query editor and visualize results with other tools such as Amazon QuickSight.

Athena is designed to make performing interactive analytics on large data collections easy. Since the service is new, we do not provide examples of how to use it for research data; we will do so in the future.

8.6 Google Cloud Datalab

The Google Cloud provides a data analytics tool called **BigQuery**, which is in many ways similar to Athena. A big difference is that while Athena is based on the standard S3 blob store, BigQuery builds on a special data warehouse. Google hosts interesting public datasets in the BigQuery warehouse, including the following.

- The names on all U.S. social security cards for births after 1879. (The table rows contain only the year of birth, state, first name, gender, and count as long as it is greater than five. No social security numbers are included.)

- New York City taxi trips from 2009 to 2015.

- All stories and comments from *Hacker News.*

- U.S. Department of Health weekly records of diseases reported from each city and state from 1888 to 2013.

- Public data from the HathiTrust and the Internet Book Archive.

- Global Summary of the Day (GSOD) weather from the National Oceanographic and Atmospheric Administration (NOAA) for 9,000 weather stations between 1929 and 2016.

Google's latest addition to BigQuery is a Jupyter-based tool called **Datalab**. (At the time of this writing, this is still a "beta" product, so we do not know what its future holds.) You can run Datalab either on their portal or on your own laptop. To run Datalab on your laptop, you need to have Docker installed. Once Docker is running and you have created a Google cloud account and created a project, you can launch Datalab with a simple docker command as illustrated in their quick-start guide [24]. When the container is up and running, you can view it at `http://localhost:8081`. What you see at first is shown in figure 8.8.

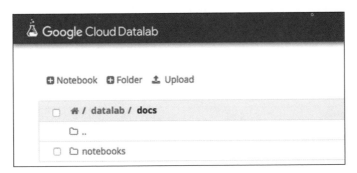

Figure 8.8: Datalab home screen. The view you see is the initial notebook hierarchy. Inside `docs` is a directory called `notebooks` that contains many great tutorials and samples.

We use two examples to illustrate the use of Datalab and the public data collections. Further details are available in notebook 15 and notebook 16.

8.6.1 Rubella in Washington and Indiana

The Google BigQuery archive contains a Centers for Disease Control and Prevention (CDC) dataset concerning diseases reported by state and city over a long period. An interesting case is rubella, a virus also known as the German measles. Today this disease has been eliminated in the U.S. through vaccination programs, except for those people who catch it in other countries where it still exists. But in the 1960s it was a major problem, with an estimated 12 million cases in the U.S. from 1964 through 1965 and a significant number of newborn deaths and birth defects. The vaccine was introduced in 1969, and by 1975 the disease was almost gone. The SQL script in figure 8.9 on the following page is based on an example in the Google BigQuery demos `cloud.google.com/bigquery`. It looks for occurrences of rubella in two states over 2,000 miles apart, Washington and Indiana, in the years 1970 and 1971. This code also illustrates the SQL "magic" operators built into Datalab, which here are used to create a callable module called `rubella`.

We can invoke this query in a Python statement that captures its result as a Pandas DataFrame and pulls apart the time stamp fields and data values:

```
rubel = bq.Query(rubella).to_dataframe()
rubelIN = rubel[rubel['cdc_reports_state']=='IN']. \
                sort_values(by=['cdc_reports_epi_week'])
rubelWA = rubel[rubel['cdc_reports_state']=='WA']. \
                sort_values(by=['cdc_reports_epi_week'])
epiweekIN = rubelIN['cdc_reports_epi_week']
epiweekWA = rubelWA['cdc_reports_epi_week']
rubelINval = rubelIN['cdc_reports_total_cases']
rubelWAval = rubelWA['cdc_reports_total_cases']
```

At this point a small adjustment must be made to the time stamps. The CDC reports times in epidemic weeks, and there are 52 weeks in a year. So the time stamp for the first and last weeks of 1970 are 197000 and 197051, respectively; the next week (the first week of 1971) is then 197100. To obtain timestamps that appear contiguous, we make a small "time slip" as follows.

```
rrealweekI = np.empty([len(epiweekIN)])
realweekI[:] = epiweekIN[:]-197000
realweekI[51:] = realweekI[51:]-48
```

Applying the same adjustment to `epiweekWA` gives us something that we can graph. Figure 8.10 shows the progress of rubella in Washington and Indiana over two years. Note that outbreaks occur at about the same time in both states and that by late 1971 the disease is nearly gone. (Continuing the plot over 1972 and 1973 shows that subsequent flare-ups diminish rapidly in size.)

```
%%sql --module rubella
SELECT   *
FROM (
  SELECT
    *,MIN(zrank) OVER (PARTITION BY cdc_reports_epi_week)AS zminrank
  FROM (
    SELECT
      *, RANK() OVER (PARTITION BY cdc_reports_state ORDER BY
            cdc_reports_epi_week ) AS zrank
    FROM (
      SELECT
        cdc_reports.epi_week AS cdc_reports_epi_week,
        cdc_reports.state AS cdc_reports_state,
        COALESCE(CAST(SUM((FLOAT(cdc_reports.cases))) AS FLOAT),0)
            AS cdc_reports_total_cases
      FROM
        [lookerdata:cdc.project_tycho_reports] AS cdc_reports
      WHERE
        (cdc_reports.disease = 'RUBELLA')
        AND (FLOOR(cdc_reports.epi_week/100) = 1970 OR
            FLOOR(cdc_reports.epi_week/100) = 1971)
        AND (cdc_reports.state = 'IN'
          OR cdc_reports.state = 'WA')
      GROUP EACH BY
        1,
        2) ww ) aa ) xx
WHERE
  zminrank <= 500
LIMIT
  30000
```

Figure 8.9: Looking for rubella in CDC reports for Indiana and Washington.

8.6.2   Looking for Weather Station Anomalies

From NOAA we have the Global Summary of the Day (GSOD) weather for 9,000 weather stations between 1929 and 2016. While not all stations were operating during that entire period, there is still a wealth of weather data here. To illustrate, we write a query to look for the hottest locations in Washington for 2015. This was a particularly warm year that brought unusual droughts and fires to the state. Our query, shown below, joins the dataset gsod2015, the table of 2015 data, with the station table to determine the station names. We order the results descending by temperature. Table 8.1 shows the top 10 results.

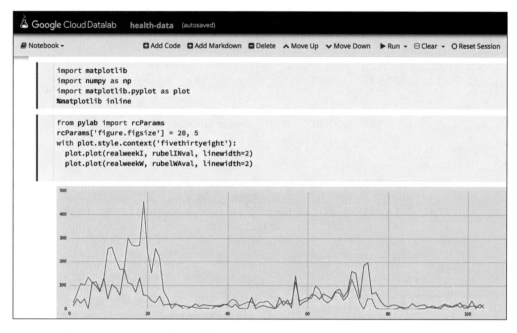

Figure 8.10: Progress of rubella in Washington (red) and Indiana (blue) from 1970 through 1971. The x axis is the week and the y axis the number of cases.

```
%%sql
SELECT maxval, (maxval-32)*5/9 celsius, mo, da, state, stn, name
FROM (
  SELECT
    maxval, mo, da, state, stn, name
  FROM
    [bigquery-public-data:noaa_gsod.gsod2015] a
  JOIN
    [bigquery-public-data:noaa_gsod.stations] b
  ON
    a.stn=b.usaf
    AND a.wban=b.wban
  WHERE
    state="WA"
    AND maxval<1000
    AND country='US' )
ORDER BY
  maxval DESC
```

Table 8.1: Hottest reported temperatures in the state of Washington in 2015.

#	Max F	Max C	Mo	Day	State	Stn	Stn name
1	113	45	06	29	WA	727846	Walla Walla Rgnl
2	113	45	06	28	WA	727846	Walla Walla Rgnl
3	111.9	44.4	06	28	WA	727827	Moses Lake/Grant Co
4	111.9	44.4	06	29	WA	727827	Moses Lake/Grant Co
5	111.2	44	09	24	WA	720272	Skagit Rgnl
6	111.2	44	09	25	WA	720272	Skagit Rgnl
7	111	43.9	06	29	WA	720845	Tri Cities
8	111	43.9	06	28	WA	720845	Tri Cities
9	111	43.9	06	27	WA	720845	Tri Cities
10	109.9	43.3	06	28	WA	720890	Omak

The results are, for the most part, what we would expect. Walla Walla, Moses Lake, and Tri-Cities are in the eastern part of the state; summer was particularly hot there in 2015. But Skagit Rgnl is in the Skagit Valley near Puget Sound. Why was it 111°F there in September? If it was so hot there, what was the weather like in the nearby locations? To find out which stations were nearby, we can look at the stations on a map. The query is simple, but it took some trial and error because the latitude and longitude for one station, SPOKANE NEXRAD, was incorrect in the database, placing it somewhere in Mongolia.

```
%%sql --module stationsx
DEFINE QUERY locations
  SELECT FLOAT(lat/1000.0) AS lat, FLOAT(lon/1000.0) as lon, name
  FROM [bigquery-public-data:noaa_gsod.stations]
  WHERE state="WA" AND name != "SPOKANE␣NEXRAD"
```

We can then invoke the Cloud Datalab mapping functions, as shown in figure 8.11 on the next page. We find that one station, called PADILLA BAY RESERVE, is only a few miles away; the next closest is BELLINGHAM INTL. We can now compare the weather for 2015 at these three locations. First, we use a simple query to get the station IDs.

```
%%sql
SELECT
  usaf, name
FROM [bigquery-public-data:noaa_gsod.stations]
WHERE
    name="BELLINGHAM␣INTL" OR name="PADILLA␣BAY␣RESERVE" \
        OR name = "SKAGIT␣RGNL"
```

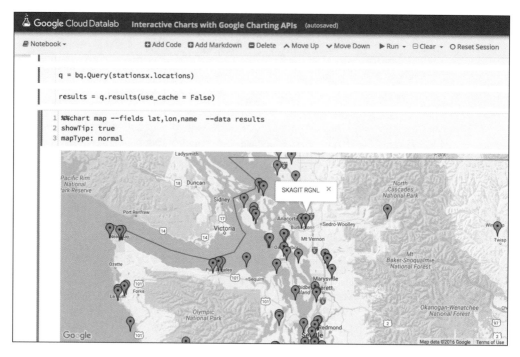

Figure 8.11: Mapping the weather stations in the northwest part of Washington.

With these results in hand, we can build a parameterized BigQuery expression:

```
qry = "SELECT max AS temperature, \
    TIMESTAMP(STRING(year) + '-' + STRING(mo) + '-' + STRING(da)) \
    AS timestamp FROM [bigquery -public -data:noaa_gsod.gsod2015] \
WHERE stn = '%s' and max <500 \
ORDER BY year DESC, mo DESC, da DESC"
stationlist = ['720272','727930', '727976']
dflist = [bq.Query(qry % station).to_dataframe() \
        for station in stationlist]
```

We can now use the following code to graph the weather for our three stations, producing the result in figure 8.12 on the following page.

```
from pylab import rcParams
rcParams['figure.figsize'] = 20,5
with plot.style.context('fivethirtyeight'):
    for df in dflist:
        plot.plot(df['timestamp'],df['temperature'],linewidth=2)
plot.show()
```

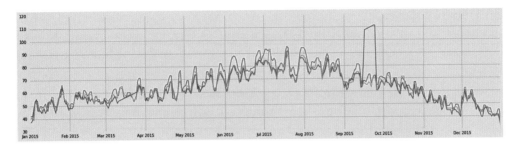

Figure 8.12: Maximum daily temperatures for Skagit (blue), Padilla Bay (red), and Bellingham (orange).

We can clearly see the anomaly for Skagit in September. We also spot another problem in March, where the instruments seemed to be not recording. Other than these, the readings are closely aligned.

These simple examples only scratch the surface of what can be done with Datalab. You can, for example, use Datalab with the TensorFlow machine learning library, and the charting capabilities are far more extensive than demonstrated here. Also, you can easily upload your own data to the warehouse for analysis.

8.7   Summary

Big data analytics may be the most widely used cloud computational capability. All companies, big or small, that run online services analyze their log data in order to understand their users and learn how to optimize their services. Moreover, the availability of large cloud-hosted scientific data collections continues to grow in fields from astronomy and cosmology to Earth science and genomics. And the original Hadoop MapReduce tools first used to perform large-scale data analytics in the cloud have now morphed into a vast suite of open source systems that are constantly being improved and extended, providing enormous power for the user but also introducing considerable complexity.

We have examined here only a few of the data analytics services available in the cloud. We introduced the architecture of YARN and then focused on Spark, which we chose because of its power and flexibility. We illustrated how Amazon Elastic MapReduce can be an excellent platform for hosting Spark. Because EMR is based on YARN, we were able to demonstrate how Azure HDInsight is similar to EMR in power. However, we did not cover the many additional services and tools that are part of the YARN ecosystem: tools such as **HBase** [138], a NoSQL database, and **Phoenix** `phoenix.apache.org`, the relational layer over HBase; **Hive**, a data warehouse and query tool; **Pig** [214], a scripting tool for orchestrating Hadoop MapReduce jobs; or **Oozie** [214], which provides workflow management for Hadoop, Pig, and Hive.

We have also taken a brief look at Amazon Athena and Azure Data Lake, services that support analysis of data in large warehouse repositories. Athena provides a portal-based tool for rapid data analysis over any data stored in Amazon S3; and Data Lake Analytics provides U-SQL, a tool for writing a graph-based parallel analysis program over petascale data in the Azure Data Lake Store. Both systems are based on serverless computing paradigms. To execute their queries, you do not need to deploy and manage VMs: the cloud infrastructure manages everything for them. We consider such issues in more detail in the next chapter.

8.8 Resources

Many excellent books cover various aspects of the topics discussed in this chapter. In particular, we recommend the following.

- *Hadoop 2 Quick-Start Guide: Learn the Essentials of Big Data Computing in the Apache Hadoop 2 Ecosystem* [114] provides a good introduction to YARN and the full Apache Hadoop ecosystem.

- *Python Data Science Handbook: Essential Tools for Working with Data* [253] provides an excellent deep dive into Python data topics, including the Pandas data analysis library and the Python machine learning library.

- *Advanced Analytics with Spark: Patterns for Learning from Data at Scale* [229] provides an in-depth discussion of Spark. We have covered Spark from the Python programmer's perspective, but Spark's native language is Scala. To appreciate the full power of Spark, you need to study this book.

Two additional Python data analysis books are *Python for Data Analysis: Data Wrangling with Pandas, NumPy, and IPython* [193] and *Python Data Analytics: Data Analysis and Science using Pandas, Matplotlib and the Python Programming Language* [209].

For those wanting to apply data analysis methods to real data, many public datasets are available to researchers in public and private clouds. The following are some datasets on the Google Cloud `cloud.google.com/public-datasets`:

- GDELT Book Corpus: 3.5 million digitized books
- Open Images Data: 9 million URLs to annotated images
- NOAA GSOD Weather Data: weather from 1929 to 2016 from 9,000 stations
- USA Disease Surveillance: reports of nationally notifiable diseases from all U.S. cities
- NYC taxi and limousine trip records from 2009 to 2015

Amazon `aws.amazon.com/public-datasets` also provides many datasets, including the following:

- Landsat and SpaceNet: Satellite imagery of all land on Earth as well as commercial satellite images
- IRS 990 tax filing from 2011 to the present
- The Common Crawl Corpus, composed of over 5 billion web pages
- NEXRAD weather radar data
- Global Database of Events, Language, and Tone (GDELT) records monitoring the world's broadcast, print, and web news

The private Open Science Data Cloud `opensciencedatacloud.org/publicdata` also has data available to researchers, including the following.

- City of Chicago public datasets
- EMDataBank: A resource for a 3-D Electron Microscopy
- FlyBase: A highly valued database for Drosophila fruit fly genomics
- General Social Survey: Answers to a range of demographic, behavioral, and attitudinal questions
- Million Song Dataset: Data and metadata for a million popular music tracks

In addition, various genomics and other life science datasets are available on all three clouds, including the following.

- The 1,000 Genomes dataset
- The Illumina Platinum Genomes, a collection of high-quality genomic datasets for community use
- The Cancer Genome Atlas (TCGA), a collection of cancer genomics data
- The 3,000 Rice Genomes collection from 89 countries.

Chapter 9

Streaming Data to the Cloud

"Panta rhei [everything flows]."

—Heraclitus

While batch analysis of big data collections is important, real-time or near-real-time analysis of data is becoming increasingly critical. Consider, for example, data from instruments that control complex systems such as the sensors onboard an autonomous vehicle or an energy power grid: here, the data analysis is critical to driving the system. In some cases, the value of the results diminishes rapidly as they get older. For example, trending topics and hashtags in a twitter stream may be uninteresting after the fact. In other cases, such as some large science experiments, the volume of data that arrives each second is so large that it cannot be retained and real-time analysis or data reduction is the only way to handle it.

We refer to the activity of analyzing data coming from unbounded streams as **data stream analytics**. While many people think this is a new topic, it dates back to basic research on complex event processing in the 1990s at places including Stanford, Caltech, and Cambridge [87]. That research created part of the intellectual foundation for today's systems. In the paragraphs that follow, we describe some recent approaches to data stream analytics from the open source community and public cloud providers. These approaches include Spark Streaming `spark.apache.org/streaming`, derived from the Spark parallel data analysis system described in the next chapter; Twitter's Storm system `storm.apache.org`, redesigned by Twitter as Heron [173]; Apache Flink `flink.apache.org` from the German Stratosphere project; and Google's Cloud Dataflow [58], becoming Apache Beam `beam.apache.org`, which runs on top of Flink, Spark, and Google's Cloud.

University projects include Borealis from Brandeis, Brown, and MIT, and the Neptune and Granules projects at Colorado State. Other commercially developed systems include Amazon Kinesis `aws.amazon.com/kinesis`, Azure Event Hubs [31], and IBM Stream Analytics [25].

The analysis of instrument data streams sometimes needs to move closer to the source. Tools are emerging to perform *preanalysis*, with the goal of identifying the data subsets that should be sent to the cloud for deeper analysis. For example, the Apache Edgent edge-analytics tools `edgent.apache.org` are designed to run in small systems such as the Raspberry Pi. Kamburugamuve and Fox [165] survey many of these stream-processing technologies, covering issues not discussed here.

In the rest of this chapter, we use examples to motivate the need for data stream analytics in science, discuss challenges that data stream analytics systems must address, and describe the features and illustrate the use of a range of open source and cloud provider systems.

9.1 Scientific Stream Examples

Many interesting examples of data stream processing in science have been examined in recent workshops [130]. We review a few representative cases here.

9.1.1 Wide-area Geophysical Sensor Networks

Many geophysical phenomena, from climate change to earthquakes and inland flooding, require vast data gathering and analytics if we are to both understand the underlying scientific processes and mitigate damage to life and property. Increasingly, scientists are operating sensor networks that provide large quantities of streaming data from many geospatial locations. We give three examples here.

Studies of ground motion conducted by the Southern California Earthquake Center `scec.org` can involve thousands of sensors that record continuously for months at high sample rates [208]. One purpose for collecting such data is to improve earthquake models, in which case data delivery rates are not of great concern. Another purpose is real-time earthquake detection, in which ground movement data are collected, analyzed to determine likely location and hazard level, and alerts generated which may, for example, be used to stop trains [174]. In this case, delivery and processing speed are vital.

The Geodesy Advancing Geosciences and EarthScope (GAGE) Global Positioning System (GPS) [220] network manages data streams from approximately

2,000 continuously operating GPS receivers spanning an area covering the Arctic, North America, and the Caribbean. These data are used for studies of seismic, hydrological, and other phenomena, and if available with low latency can also be used for earthquake and tsunami warnings.

The U.S. National Science Foundation-funded **Ocean Observatories Initiative** (OOI) [102, 34] operates an integrated set of science-driven platforms and sensor systems at multiple locations worldwide, including cabled seafloor devices, tethered buoys, and mobile assets. 1,227 instruments of 75 different types collect more than 200 different kinds of data regarding physical, chemical, geological, and biological properties and processes, from the seafloor to the air-sea interface. Once acquired, raw data (consisting mostly of tables of raw instrument values—counts, volts, etc.) are transmitted to one of three operations centers and hence to mirrored data repositories at east and west coast cyberinfrastructure sites, where numerous other derived data are computed. The resulting data are used to study issues such as climate change, ecosystem variability, ocean acidification, and carbon cycling. Rapid data delivery is important because scientists want to use near-real-time data to detect and monitor events as they are happening.

9.1.2   Urban Informatics

Cities are complex, dynamic systems that are growing rapidly and becoming increasingly dense. The global urban population in 2014 accounted for 54% of the total global population, up from 34% in 1960. In the U.S., 62.7% of the people live in cities, although cities constitute only 3.5% of the land area [98]. Cities consume large quantities of water and power, and contribute significantly to greenhouse gas emissions. Making urban centers safe, healthy, sustainable, and efficient is of critical global importance.

Understanding how cities work and how they respond to changing environmental conditions is now part of the emerging discipline called **urban informatics**. This new discipline brings together data analytics experts, sociologists, economists, environmental health specialists, city planners, and safety experts. In order to capture and understand the dynamics of a city, many municipalities have begun to install instrumentation that helps monitor energy use, air and water quality, the transportation system, crime rates, and weather conditions. The aim of this real-time data gathering and analysis is to help municipalities avert citywide or neighborhood crises and more intelligently plan future expansion.

Array of Things. A project in the city of Chicago called the **Array of Things** `arrayofthings.github.io` is leading the way in defining how data can be collected and analyzed from instruments placed within cities. The project is led by Charlie Catlett and Peter Beckman from the Urban Center for Computation and Data, a joint initiative of the University of Chicago and Argonne National Laboratory. The Array of Things team has designed the hardware and software for a sensor pod that can be placed on utility poles throughout a city, gather local data, and push the data back as a stream to a recording and analysis point in the cloud. As shown in figure 9.1 on the following page, the sensor package contains numerous instruments, a data-processing engine, and a communications package. Because the sensor package is intended to be installed on poles and infrequently accessed, reliability is an important feature. Consequently, the system contains a sophisticated reliability management subsystem that monitors the instruments, computing, and communications, and reboots failed systems if needed and possible.

Sensors in the sensor pod can measure temperature and humidity; physical shock and vibration; magnetic fields; infrared, visible, ultraviolet light; sound; and atmospheric contents such as carbon monoxide, hydrogen sulphide, nitrogen dioxide, ozone, sulfur dioxide, and air particles. The pod also contains a camera, but it cannot transmit identifiable images of individual people. The data from these instruments are uploaded approximately every 30 minutes as a JSON record, which approximately takes the form shown below.

```
{
    "NodeID":     string,
    "timestamp":  2016-11-10 00:00:20.143000,
    "sensor":     string,
    "parameter":  string,
    "value":      string
}
```

Note that a data stream from the pod may have more than one sensor and that individual sensors can take multiple different measurements. These measurements are distinguished by the `parameter` keyword. We use this structure in an example later in this chapter.

9.1.3 Large-scale Science Data Flows

At the other end of the application spectrum, massively parallel simulation models running on supercomputers can generate vast amounts of data. Every few simulated time steps, the program may generate large (50 GB or more) datasets distributed over thousands of processing elements. While you can save a few such datasets to

Figure 9.1: Array of Things sensor pod contents (top) and pole mount (bottom.)

a file system, it is often preferable to create a stream of these big objects and let another analysis system consume them online, without writing them to disk.

The **ADIOS** [184] HPC I/O library supports this mode of interaction. It provides a simple and uniform API to the application programmer, while allowing the backend to be adapted to a variety of storage or networking layers while taking full advantage of the parallel I/O capabilities of the host system. One such backend leverages a networking layer, EVPath [115], that provides the flow and control needed to handle such a massive stream. Another backend target for ADIOS is DataSpaces [112], a system for creating shared data structures between applications across distributed systems. DataSpaces accomplishes this by mapping n-dimensional array objects to one dimension by using a distributed hash table

165

and Hilbert space filling curves. Together, these components provide a variety of streaming abstractions that can be used to move data from a HPC application to a range of HPC data analysis and visualization tools.

9.2 Basic Design Challenges of Streaming Systems

Designers of streaming systems face a number of basic challenges, including correctness and consistency. Data in an unbounded stream are unbounded in time. But if you want to present results from the analytics, you cannot wait until the end of time. So instead you present results at the end of a reasonable time window. For example, you may specify a daily summary based on a complete checkpoint of events for that day. But what if you want results more frequently, say, every second? If the processing is distributed and the window of time is short, you may not have a way to know the global state of the system, and some events may be missed or counted twice. In this case the reports may not be consistent.

Strongly consistent event systems guarantee that each event is processed once and only once. In contrast, a **weakly consistent** system may give you only approximate results. You can verify these results, if needed, by performing a batch run on each daily checkpoint file, but this verification step involves additional work and delays. Streaming system designs that combine a streaming engine with a separate batch system are examples of lambda architecture [190]. The goal of many of the systems described below is to combine the batch computing capability with the streaming semantics so that a separate batch system is not necessary.

A second concern is the semantics of time and windows. Many event sources provide a time stamp when an event is created and pushed into the stream. However, the event will not be processed until a later time. Thus, we need to distinguish between event time and processing time. To further complicate things, events may be processed out of event-time order. These factors raise the question of how to reason about event time in windows defined by processing time.

At least four types of time windows exist. **Fixed time** windows divide the incoming stream into logical segments, each corresponding to a specified interval of processing time. The intervals do not overlap. **Sliding** windows allow for the windows to overlap: for example, windows of size 10 seconds that start every five seconds. **Per session** windows divide the stream into sessions of activity related to some key in the data. For example, mouse clicks from a particular user may be bundled into a sequence of sessions of clicks nearby in time. **Global** windows can encapsulate an entire bounded stream. Associated with windows there must be a mechanism to trigger an analysis of the content of the window and publish the summary. Each of the systems that we discuss below support some windowing mechanisms. Two articles by Tyler Akidau [57] provide a good discussion of windows and related issues.

Another design issue concerns how work is distributed over processors or containers and how parallelism is achieved. As we will see, the systems described here all adopt similar approaches to parallelism.

Operations on streams often resemble SQL-like relational operators, but there are important differences. In particular, a join operation is not well defined when streams are unbounded. The natural solution involves dividing streams by windows in time and performing the join over each window. Vijayakumar and Plale have studied this topic extensively [255]. Barja et al. [66] describe a Complex Event Detection and Response system in which SQL-like temporal queries have well-defined semantics.

9.3   Amazon Kinesis and Firehose

Amazon provides an impressive event-streaming software stack called **Kinesis**, comprising the following three services:

1. **Kinesis Streams** provides ordered, replayable real-time streaming data.

2. **Kinesis Firehose**, designed for extreme scale, can load data directly into S3 or other Amazon services.

3. **Kinesis Analytics** provides SQL-based tools for real-time analysis of streaming data from Kinesis Streams or Firehose.

9.3.1   Kinesis Streams Architecture

Each Kinesis stream is composed of one or more *shards*. Think of a stream as a rope composed of many strands. Each strand is a shard, and the data that move through the stream are spread across the individual shards that make up the stream. Data producers write to the shards and consumers read from the shards. Each shard can support writes from producers at up to 1,000 records per second, up to a maximum data write total of 1 MB per second. However, no individual record can be bigger than 1 MB. On the other hand, reads by data consumers can be at most five transactions per second, with a total throughput of 2 MB/sec. While these bounds may seem rather limiting, you can have streams with thousands of shards. For scientists, the biggest limitation may be the 1 MB limit on the size of each event, which means that big events must split up and spread over multiple shards. We return to this detail later.

To create a stream, go to the Amazon console, click on the word $\boxed{\text{stream}}$, give the stream a name, and select the desired number of shards, as shown in figure 9.2.

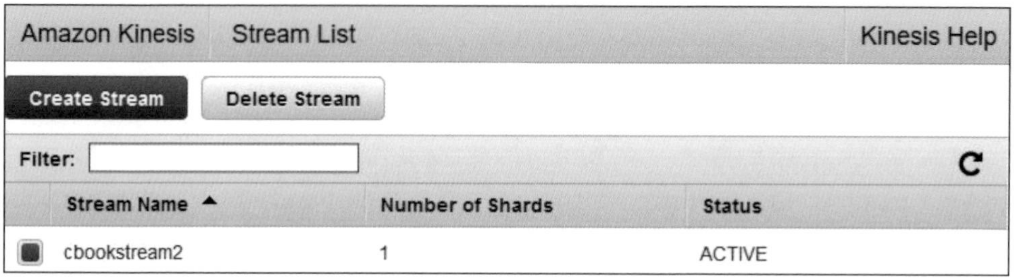

Amazon Kinesis	Stream List		Kinesis Help

Create Stream	Delete Stream

Filter: [   ]   **C**

Stream Name ▲	Number of Shards	Status
■ cbookstream2	1	ACTIVE

Figure 9.2: Creating a single stream with one shard from the Amazon console.

Let's start with an example of sending a record to the stream `cbookstream2`. Assume we want to send a series of time-stamped temperature readings from a temperature sensor. Every record has to have a stream name, a binary encoded data component, and a way to identify the shard. The Boto3 SDK that we use here provides an interface that is consistent with that used for other Amazon services. We identify the shard by giving the record a partition key as a string: in this case, as we have just one shard, the string `'a'`. This key is hashed to an integer that is used to select a shard. (If there is only one shard, all records are mapped to shard 0.) In our example, the binary data component of our record is up to us. We create a string that encodes a JSON record and then convert the string to a binary array. The following code sends our record to the stream. (We use the `datetime` format for the timestamp because that is the format used by Streams for its timestamps.)

```
client = boto3.client('kinesis')
tz = pytz.timezone('America/Los_Angeles')
ts = datetime.datetime.now(tz)
item = {'id': 'sensor 1', 'val': 73, 'label': 'temperature',
        'localtime': str(ts)}
data = json.dumps(item)
client.put_record(
        StreamName='cbookstream2',
        Data= bytearray(data),
        PartitionKey = 'a'
    )
```

Reading a stream takes a little more work. Every record loaded into a shard has a sequence number. To read the records, you need to provide a *shard iterator*, which can be created in several ways. One way is to specify a timestamp so that you read only the records that arrive after the specified time. Another way is to request that the iterator be positioned to be the latest point in the stream so that you only get the new records that come after that point. Or you can create an iterator from a sequence number. To do so, you also need the shardID. You can obtain both the shardID and a starting sequence number for that shard by calling `describe_stream(StreamName)`. Given this information, you can create an iterator as follows.

```
client = boto3.client('kinesis')
iter = client.get_shard_iterator(
     StreamName='cbookstream2',
     ShardId='shardID',
     ShardIteratorType = 'AT_SEQUENCE_NUMBER',
     StartingSequenceNumber='seqno'
     )
```

If you want an iterator that starts only after the latest record, so as to obtain only new records, you can set the `ShardIteratorType` equal to "LATEST" and omit the sequence number. (Two other iterator constructor methods also can be used, but we do not discuss them here.)

Having created an iterator, we can then ask for all records collected from that point on by using the `get_records()` function. This function returns at most 10 MB in one call and can support only 2 MB per second. To avoid the 10 MB limit, you can limit the number of records returned; but if you are near the limit, it is better to add new shards or, alternatively, to use a split-shard function.

The best strategy is to pull records in time-defined windows and then do the analysis for each window. The `get_records()` function returns a list of records and metadata that includes a "next shard iterator" that can be used to position the function for the next batch of records. For example, taking the result from the previous code, a typical processing loop would look like the following.

```
iterator = iter['ShardIterator']
while True:
    time.sleep(5.0)
    resp = client.get_records(ShardIterator=iterator)
    iterator = resp['NextShardIterator']
    analyzeData(resp['Records'])
```

This code pulls a block of records that have been waiting in the shard for the last five seconds (not counting the time to do the data analysis). For example, suppose we want to measure the time that elapses from when the stream creates a record, as described above, until the time the record arrives and is timestamped in the stream. We could write it as follows.

```
def analyzeData(resp):
    #resp is the response['Records'] field
    for rec in resp:
        data = rec['Data']
        arrivetime = rec['ApproximateArrivalTimestamp']
        print('Arrival time = '+str(arrivetime))
        item = json.loads(data)
        prints('Local time = ' + str(parse(item['localtime'])))
        delay = arrivetime - parse(item['localtime'])
        secs = delay.total_seconds()
        print('Message delay to stream = '+str(secs) + ' seconds')
```

9.3.2 Kinesis and Amazon SQS

It is instructive to compare Kinesis Streams and Kinesis Firehose with the Amazon **Simple Queue Service** (SQS) introduced in section 2.3.6 on page 34. SQS is based on the semantics of queues. Message (event) producers add items to SQS queues; SQS clients can retrieve messages from queues for processing. Messages in a queue are not removed by clients, but are instead retained for a period of time (typically 24 hours) or until the entire queue is explicitly flushed. Each message in a queue has a sequence number; knowing this number, the client can fetch that message and all subsequent messages (up to a limit) with a single call to the stream API. Thus, a client can replay an analysis of a queue at any time, and different clients can process the same queue in the same or different ways.

In contrast, Kinesis Firehose is designed to handle large streams of data that are automatically delivered to S3 or Amazon Redshift. Firehose is batch oriented: it buffers incoming data into buffers of up to 128 MB and dumps the buffer to your S3 blobs at specific intervals that you select, from every minute to every 15 minutes. You can also specify that the data be compressed and/or encrypted. Consequently Firehose is not designed for real-time analysis but for near-real-time large-scale analysis.

9.4 Kinesis, Spark, and the Array of Things

To illustrate the use of Kinesis together with Spark in a streaming context, we study the data from 40 different instrument streams coming through Kinesis. Spark has a streaming subsystem called **Spark Streaming**. The key concept is that of a **D-stream**, which takes windows of streamed data and blocks them into Spark RDDs for analysis. We use these concepts here to implement a basic anomaly prediction algorithm. More specifically, we build a Jupyter notebook that tracks all the streams but focuses on two where sudden dramatic changes in behavior have been seen. Our goal is a system that automatically flags these anomalies. To make this easier to do in a notebook, we use regular Spark plus Kinesis, rather than the Spark streaming subsystem. We create a pseudo D-Stream by pulling blocks of events from Kinesis and converting each block into a Spark RDD.

The data that we work with here are based on a 24-hour sample of event streams from 40 instruments deployed as part of the Array of Things project described in section 9.1.2 on page 163. We make these data available for download from the book website; many more are available from `arrayofthings.github.io`.

We develop a program to read the dataset and push the events as fast as possible to Kinesis. While we do not present all code details here—we want to focus on the core ideas—the details are in notebook 17. (The book website has links to the code and data for this example in the **EXTRAS** tab.) We first create an instance of Jupyter running with Spark. (The example can run on a laptop.) We connect to Kinesis exactly as described in the previous section and start pulling event records and accumulating them into RDD windows.

We use a simple algorithm to detect anomalies: We record the data stream values for each stream of interest; and as we are recording them, we compute a simple predicted value for the next value. If the next value from the stream turns out to be extremely different from our prediction, then we flag an anomaly.

To record the data streams, we create an instance of a class `Datarecorder`. This has one main method: `record_newlist(newlist)`, which takes a list of event records of the form (timestamp, value) and appends it to `self.datalist`, which is the record of the stream. In addition to this record, we use the recent history of the stream to signal unexpected changes in the stream, such as anomalies or other major changes in behavior. A challenge is that some of the signals can be noisy, so we must look for ways to filter the noise to see the significant changes in behavior.

To filter the noise in a stream $x_i, i = 0...$, we can use an exponential-based smoothing technique that blends the previous values from the stream to create an artificial stream s_i that tends to a geometric average of past values as follows. Let $s_0 = x_0$, and use the following recurrence to define the future values:

$$s_n = (1 - \alpha)x_n + \alpha s_{n-1},$$

where α is a number between 0 and 1. Expanding the recurrence, we see that:

171

$$s_n = (1 - \alpha)x_n + \alpha(1 - \alpha)x_{n-1} + \alpha^2 s_{n-2}$$

$$s_n = (1 - \alpha) \sum_{i=0}^{n-1} \alpha^i x_{n-i} + \alpha^n x_0$$

(One can easily verify that in the case of a constant stream with $x_i = x_0$ for all i, then $s_i = x_i$ for all i.) If the stream makes a radical change in value, however, the smoothed value lags behind. We can use this lag to identify anomalies and other points of behavior change. But the problem then becomes how to measure a profound change in behavior in a manner that is distinct from noise. To do that, we can compute the standard deviation and look for departures from our smoothed value that exceed the standard deviation.

We cannot readily compute the standard deviation of stream values that may dramatically change over time, but we can compute the standard deviation in a recent window. We create a buffer, buf_i, $i = 1..M$, to record the most recent M stream values. We can compute the standard deviation σ in this window as follows.

$$\mu = \frac{1}{M} \sum_{i=1}^{M} buf_i$$

$$\sigma = \sqrt{\frac{1}{M} \sum_{i=1}^{M} (buf_i - \mu)^2}$$

Based on this computation, we can look for values of x_i that lay outside the interval $[s_i - k\sigma, s_i + k\sigma]$, where k is some value greater than 1. We use $k = 1.3$ here. While 2σ would show us some truly significant outliers, we have found that the more modest 1.3σ worked well. This computation takes place in the Datarecorder `record_newlist(newlist)` method. We also keep a record of s_i, $s_i - k\sigma$ and $s_i + k\sigma$ that we can plot after we complete the analysis.

We look at two of the 40 streams by creating a dictionary of the data recorders for each stream of interest. One is a chemistry sensor that tracks atmospheric no_2, a common pollutant; and the other is an ambient temperature sensor. We create a function `update_recorders(newlist)`, shown on the next page, which takes a list of the following form, and uses the dictionary to select the correct recorder, passing the timestamp and value list to the recorder function.

```
[ [sensor-name, [[time-stamp, value], [time-stamp, value] .... ]
  [sensor-name, [[time-stamp, value], [time-stamp, value] .... ]
    ..
]
```

```
myrecorder = {}
myrecorder['Chemsense_no2'] = Datarecorder('Chemsense_no2')
myrecorder['TSYS01_temperature'] = Datarecorder('TSYS01_temperature')

def update_recorders(newlist):
    if newlist != None and newlist != []:
        for x in newlist:
            myrecorder[x[0]].record_newlist(x[1])
```

Our Spark-Kinesis pipeline has four stages. The first stage creates the RDD in our pseudo D-Stream:

- `gather_list(iteratorlist)` takes the list of Kinesis stream iterators and pulls all the events since the last call to this function from the stream. It then updates `iteratorlist` with the iterator advanced to the spot that follows the last one we pulled. Each event is a binary-encoded JSON object so it is converted into a full JSON object and then to a list.

- `filter_fun(row, sensor, parameter)` is used to select the elements of the RDD that correspond to a specific sensor, parameter pair.

The main loop of the program emulates Spark streaming. Approximately every 20 seconds, we gather all available events from the Kinesis stream and then create a Spark RDD for them called `data`. Each event is a list of the following form.

```
[sensor-name, [timestamp, value]]
```

The first step in the pipeline filters all events except those that we want to keep. The second step groups the events by the sensor-name key in the tuple, producing a list with two elements, as follows.

```
[ ['Chemsense-no2', [python-iterator over (time-stamp, value) tuples]],
  ['TSY01-temperature', [python-iterator over(time-stamp, value) tuples]]
]
```

The third step uses `map` to convert the Python iterators into explicit lists, using a simple function `doiter(row)`. We then `collect` these lists into a single `newlist`, which we pass to the recorders to record and look for events of interest.

```
for i in range(150):
    gathered = gather_list(iterlist)
    data = sc.parallelize(gathered, 2)
    newlist = data.filter(lambda p: filter_fun(p,'Chemsense','no2')
                        or filter_fun(p,'TSYS01','temperature')) \
                .groupByKey()  \
                .map(lambda p: doiter(p))   \
                .collect()
    update_recorders(newlist)
    print('*********   end of gather %s ***************'%i)
    time.sleep(20.0)
```

Note that the data in the dataset cover 24 hours of real data, during which time the instruments send an event approximately every 25 seconds: approximately two events per minute for a total of 3,450 for the full day. Altogether there are 172,800 events in the dataset, and the total size is approximately 14 MB. We push all data to Kinesis in about 120*20 seconds = 40 minutes using one shard. We could do it much faster with two shards.

The output is rather dull until we get to timestamp 17:05:06, when we see the following anomalies reported.

```
[updating list for TSYS01_temperature
********* end of gather 84 **************
updating list for Chemsense_no2
anomaly at time 2016-11-10 17:05:06.980000?
anomaly at time 2016-11-10 17:05:57.049000
... lines deleted ...
anomaly at time 2016-11-10 17:13:26.838000
anomaly at time 2016-11-10 17:13:51.875000
updating list for TSYS01_temperature
anomaly at time 2016-11-10 17:15:57.064000
... lines deleted ...
anomaly at time 2016-11-10 17:29:17.405000
anomaly at time 2016-11-10 17:29:42.460000
anomaly at time 2016-11-10 17:30:07.507000
... more time passes ...
updating list for TSYS01_temperature
anomaly at time 2016-11-10 22:23:07.533000
anomaly at time 2016-11-10 22:23:32.592000
anomaly at time 2016-11-10 22:24:22.673000
... ending at ...
anomaly at time 2016-11-10 23:14:47.974000
anomaly at time 2016-11-10 23:17:18.233000
********* end of gather 114 **************
```

Each recorder keeps track of the data, smoothed prediction, and 3σ-wide safety window. The plot function on the Datarecorder shows the history. We can clearly see the periods of strange behavior, as shown in figure 9.3.

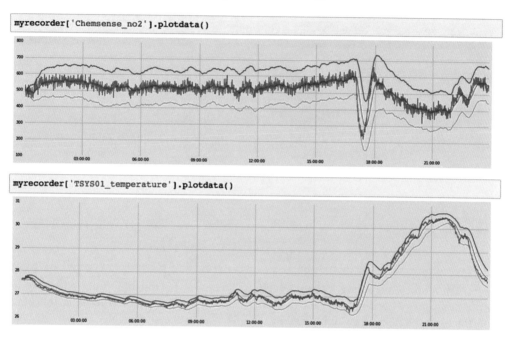

Figure 9.3: Plot of the data streams from the sensors. The high frequency blue line is the raw data. The red line is the smoothed prediction line. The other two lines show 1.5σ above and below the prediction, respectively. When the blue line escapes the 3σ window, the anomaly is signaled.

9.5 Streaming Data with Azure

Azure has a set of services that are devoted to real-time, large scale stream analytics. The system is designed to handle millions of events per second and to correlate across multiple streams of data. Internet of Things applications are particularly well handled here. The two primary components of the Azure stream analytics services are the **Azure Event Hubs** service and **Stream Analytics** engine.

Event Hubs is where your instruments send their events. It is similar in this respect to Kinesis. The analytics portal is where you put your event processing logic, which is expressed as a streaming dialect of SQL. As illustrated in figure 9.4 on the next page, the Stream Analytics portal allows you to select an input stream from an Event Hub or from blob storage and use it as input to the query system. You can then direct the output to blob storage. As illustrated in the figure, the output of the query may go directly to the portal console.

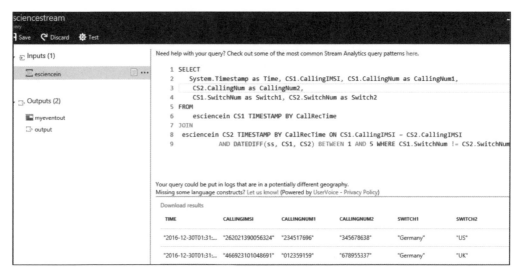

Figure 9.4: View of an Azure stream analytics query. In this example we are looking for telephone fraud. The input stream is a set of call logs. The query is looking for a call from the same number from different locations at nearly the same time.

The best way to create an Event Hub is via the Azure portal. Once that is accomplished, sending events to the Event Hub is easy. The Event Hubs service identifies hub instances by a *name space*, an *Event Hub name*, an *access key name*, and an *access key*. These values are all available from the portal. You can create as many hub instances in a namespace as you need. The maximum size of an event is 256 KB, so if your data are larger, you must break them into 256 KB blocks and send them in batches. The Event Hub has a sequence number so that you can keep track of the individual parts of your big message.

Suppose your events are scientific records with the form `category, title, abstract`. You set up a connection to the Event Hub and send an event triple `doc` as follows, labeling each element of the event triple appropriately.

```
from azure.servicebus import ServiceBusService

servns = 'myhubnamespace'
key_name = 'RootManageSharedAccessKey'  # default from Azure portal
key_value = 'longkey from portal'
sbs = ServiceBusService(service_namespace=servns,
                        shared_access_key_name=key_name,
                        shared_access_key_value=key_value)

event_data = {'category':doc[0], 'title':doc[1], 'abstract':doc[2]}
sbs.send_event('event hub name', json.dumps(event_data))
```

Events from the Event Hub are typically routed to the Azure Stream Analytics engine. As shown in figure 9.5, you can save the output from the engine to blob storage; route it to the Azure Queue Services, where your own microservices can subscribe to events and process them; or have the Stream Analytics service queries invoke the Azure Machine Learning services.

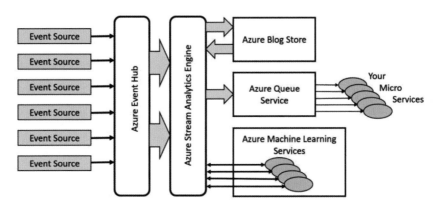

Figure 9.5: Events move from sources to the Event Hub and then to the Stream Analytics engine. From there they may go to blob storage, the Azure queue service, or the Azure Machine Learning service.

To show how the machine learning system works with stream analytics, we need to look more closely at the SQL language dialect that is part of the analytics services. Language extensions have been added to accommodate a library of mathematical and logical functions, as well as windowing semantics. Similar concepts are part of the query language used by Amazon Kinesis Analytics. For example, suppose we have a stream of temperature reading events, each containing a sensor name and a temperature value, and you want to know the standard deviation of the temperature in a window of 20 minutes for each sensor. We can ask this question as follows. The `GROUPBY` operator gathers all the events in the window with the same sensor name, and the `STDEV` operator computes the standard deviation for each sensor's temperature values. The term **tumbling window** indicates that windows are fixed size, non-overlapping, and contiguous.

```
SELECT sensorname, STDEV(temperature)
FROM input
GROUPBY sensorname, TumblingWindow(minute, 20)
```

Azure Stream Analytics allows us to invoke a machine learning function in the same way. For example, assume that we have an Azure Machine Learning (ML) web service that can take the abstract for a scientific document and predict a pair of classifications: a best guess and a second best guess. We can now take the stream of scientific document events and directly invoke the classifier with the following Stream Analytics query.

```
WITH subquery AS {
    SELECT category, title, classify(category, title, abstract)
    as result from streaminput
 }
SELECT category, result.[Scored Label], result.[second], title
FROM
    subquery
INTO
    myblobstore
```

The `classify()` web service ignores the category and title and uses machine learning libraries to analyze the text of the abstract and returns a result with two fields, `Scored Label` and the second-choice score `second`. The result is stored as a CSV file in an Azure blob. In the next chapter we describe how Azure ML can be used to construct such a web service.

You might well wonder how well Azure Stream Analytics can keep up with the flow of input events, given that it must invoke a separate web service for each query evaluation. The short answer is "very well." To demonstrate, we sent a batch of 1,939 events to an Event Hub from a Jupyter notebook. Table 9.1 presents the total time taken to process the entire batch. We also measure the time taken to invoke the Azure ML service directly for all 1,939 events, with one invocation per event. We also measure the time taken when we invoke the Azure ML service with a bulk request, whereby all events are sent in one web service call.

We see that bulk processing of events can have a substantial impact on performance. When requests are sent one at a time (the second row in the table), 1,939 events take 1,042 seconds, or approximately 0.5 seconds to make a single call to

Table 9.1: Time taken by Azure Stream Analytics to process 1,939 input events.

Metric	Time (seconds)
Total time for Azure Event Hub invocation	162
Total time for direct Azure ML invocations	1,042
Total time for bulk Azure ML invocation	3.75
Average time between event arrivals to Event Hub	0.0825
Average time for web service call in Stream Analytics	0.0827

the Azure ML web service and return the result. However, if we send the events in one large block request (the third row), it takes only 3.75 seconds, an average of 0.002 seconds per event. As the Stream Analytics system automatically groups requests to the web service, it performs only 578 distinct invocations for the 1,939 events, as revealed by the automatic monitoring shown in figure 9.6. This grouping allows it to process each event in an average of 0.0827 seconds (the fifth line in the table), which matches the arrival rate of events (the fourth line).

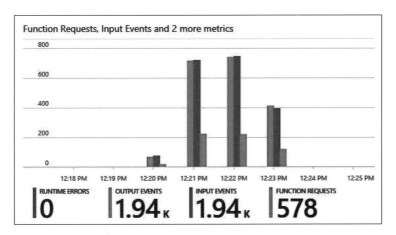

Figure 9.6: Metrics captured for our experiment by Azure Stream Analytics. Each set of three bars shows in turn output events, input events, and web service invocations.

We also experiment with doubling the event arrival rate by using two machines to send the complete set of events simultaneously, and saw no significant impact on performance. The Stream Analytics system kept up and required only 598 Azure ML function invocations, even though twice as many events were processed. You might want to try an interesting exercise: use many input event sources to test the limits of the Azure Stream Analytics system's scalability.

9.6 Kafka, Storm, and Heron Streams

Apache **Kafka** `kafka.apache.org` is an open source message system that incorporates publish-subscribe messaging and streaming. It is designed to run on a cluster of servers and is highly scalable. Streams in Kafka are streams of records that are segregated into *topics*. Each record has a key, a value, and a timestamp.

Kafka streaming can be simple—a single-stream client consuming events from one or more topics—or more complex, based on sets of producers and consumers organized into graphs, called **topologies**. We do not go into the details of Kafka here. Instead, we provide a brief overview of two systems that are similarly based on executing a directed graph of tasks in a dataflow style.

One of the earliest such systems was Storm, created by Nathan Marz and released as open source by Twitter in late 2011. Storm was written in a dialect of Lisp called Clojure that runs on the Java virtual machine. In 2015, Twitter rewrote Storm to create the API-compatible Heron [173] `twitter.github.io/heron`. Since Heron implements the same programming model and API as Storm, we discuss Storm first and then say a few words about the Heron design.

Storm (as well as Heron) runs *topologies*, directed acyclic graphs whose nodes are **spouts** (data sources) and **bolts** (data transformation and processing). Figure 9.7 shows an example Storm topology.

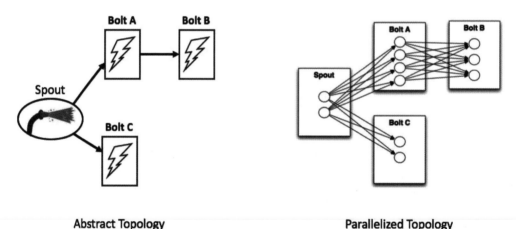

Figure 9.7: An example Storm topology. On the left is the abstract topology as defined by the program; on the right is the unrolled parallel topology as executed at runtime.

Storm has two programming models: *classic* and *Trident*, with the latter built on the first. When using the Storm programming model, you express your application logic by extending the basic spout and bolt classes and then using a topology builder to tie everything together. Figure 9.8 shows a basic template for a bolt, expressed in Java as that is Storm's main programming language interface. Three methods are required: `prepare()`, `execute()`, and `declareOutputFields()`.

The `prepare()` method is a special constructor that is called when the actual instance is deployed on the remote machine. It is supplied with context about the configuration and topology as well as a special object called the OutputCollector, which is used to connect the bolt output to the output stream defined by the topology. This method is used to instantiate your own data structures.

The basic data model for Storm/Heron is a stream of tuples. A tuple is just that: a set of items, each of which needs only to be serializable. Some fields in a tuple have names that are used for communicating information between bolts. The `declareOutputFields()` method is used to declare the name of the fields in a stream. We discuss this method further later.

The heart of the bolt is the `execute()` method, which is invoked once for each new tuple that is sent to the bolt. This method contains the bolt's computational core and is also where results from the bolt process are sent to its output streams. Other classes and styles of bolts are also provided. For example, one specialized bolt class provides for sliding and tumbling windows.

Spouts are similar to classes; the most interesting are those that connect to event providers such as Kafka or Event Hubs.

Having defined a set of bolts and spouts, you develop a program by using the topology builder class to build the abstract topology and define how the parallelism should be deployed. The key methods are `setBolt()` and `setSpout()`. Each takes three arguments: the name of the spout or bolt instance, an instance of your spout or bolt class, and an integer (the *parallelism number*) that specifies how many tasks to assign to execute the instance. A task is a single thread that is assigned to a spout or bolt instance.

The code on the page after next shows how to create the topology of figure 9.7. We see that there are two tasks for the spout, four for bolt A, three for bolt B, and two for bolt C. Note that the two tasks for the spout are sent to four tasks for bolt B. We use a stream grouping function to partition the two output streams over the four tasks: specifically, shuffle grouping, which distributes them randomly. When mapping the four output streams from bolt A to the three tasks of bolt B, we use a field grouping based on a field name to ensure that all tuples with the same field name are mapped to the same task.

```
public class MyBolt extends BaseRichBolt{
    private OutputCollector collector;

    public void prepare(Map config, TopologyContext context,
                        OutputCollector collector) {
            this.collector = collector;
    }
    public void execute(Tuple tuple) {
        /*
         * Execute is called when a new tuple has been delivered.
         * Do your real work here. For example, create
         * a list of words from the tuple and then emit
         * those words to the default output stream.
         */
            for(String word : words){
                    this.collector.emit(new Values(word));
            }
    }
    public void declareOutputFields(OutputFieldsDeclarer declarer)
        {
        /*
         * The declarer is how we declare our output fields in
         * the default output stream. You can have more than
         * one output stream, using declarestream. The emit()
         * in execute needs to identify the stream for each
         * output value.
         */
        declarer.declare(new Fields("word"));
    }
}
```

Figure 9.8: A basic Store/Heron bolt template, with three methods: prepare(), execute(), and declareOutputFields().

```
TopologyBuilder builder = new TopologyBuilder();
builder.setSpout("Spout", new MySpout(), 2);
builder.setBolt("BoltA", new MyBoltA(), 4).shuffleGrouping("spout");
builder.setBolt("BoltB", new MyBoltB(), 3).fieldsGrouping("BoltA",
                        new Fields("word"));
builder.setBolt("BoltC", new MyBoltC(), 2).shuffelGrouping("spout")

Config config = new Config();
LocalCluster cluster = new LocalCluster();
cluster.submitTopology("mytopology", config,
                                builder.createTopology());
```

Having described the Storm/Heron API, we provide some information on the Heron implementation. In Heron, a set of container instances is deployed to manage the execution of the topology, as shown in figure 9.9. The topology master coordinates the execution of the topology on a set of other containers each of which contains a stream manager and Heron instance processes that execute the tasks for the bolts and spouts. The communication between the bolts and spouts is mediated by the stream manager, and all the stream managers are connected together in an overlay network. (The topology master makes sure they are all in communication.) Heron provides considerable performance improvements over Storm. One improvement is better flow control of data from spouts when the bolts are falling behind. For more detail on Heron, see the full paper [173]. Some of the best Storm tutorial materials can be found in Michael Noll's blog `www.michael-noll.com`.

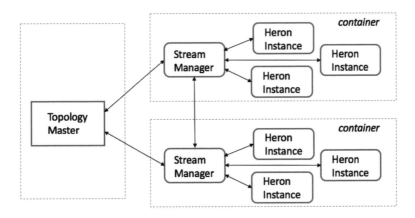

Figure 9.9: Heron architecture detail.

9.7 Google Dataflow and Apache Beam

Apache Beam `beam.apache.org`, the open source release of the Google Cloud Dataflow system [58], is the most recent entry to the zoo of data stream analytics solutions that we discuss here. An important motivation for Beam (we write Beam rather than Google Cloud Dataflow from now on, to save space) is to treat the batch and streaming cases uniformly. The important concepts are as follows.

1. **Pipelines**, which encapsulate computation

2. **PCollections**, which represent data as they move through a pipeline

3. **Transforms**, the computational transformations that operate on PCollections and produce PCollections

4. **Sources** and **Sinks**, from which data are read and to which data are written, respectively

A PCollection can comprise either a large but fixed-size set of elements or a potentially unbounded stream. The elements in any PCollection are all of the same type, but that type may be any serializable Java type. The creator of a PCollection often appends a timestamp to each element at creation time, particularly when dealing with unbounded collections. One important type of PCollection that is often used is the key-value PCollection, `KV<K, V>`, where `K` and `V` are the key and value types, respectively. Note that PCollections are immutable: you cannot change them, but you can apply transforms to translate them into new PCollections.

Without going into the details of how you initialize a pipeline, we discuss here how to create a PCollection of type `PCollection<String>` of strings from a file. (The API uses the Java programming language, but we hope that it is still readable by the Python programmer.)

```
Pipeline p = Pipeline.create(options);
PCollection<String> pc =
        p.apply(TextIO.Read.from("/home/me/mybigtextfile.txt"))
```

We have used the pipeline operator `apply()`, which allows us to invoke the special transform `TextIO` to read the file. Now we create a sequence of PCollections using the `apply()` method of the PCollection class. The library has five basic transform types, most of which take a built-in or user-defined function object as an argument and apply the function object to each element of the PCollection to create a new PCollection:

- **Pardo** applies the function argument to each element of the input PCollection. The computations are performed by the worker tasks allocated to this activity, in what is basic embarrassingly parallel map parallelism.

- **GroupByKey**, when applied to a KV<K,V> type of PCollection, groups all elements with the same key into a single list, so that the resulting PCollection is of type KV<K, Iterable<V> >. In other words, this is the shuffle phase of a MapReduce.

- **Combine** applies an operation that reduces a PCollection to a PCollection with a single element. If the PCollection is windowed, the result is a PCollection with the combined result for each window. Another type of combining is for key-grouped PCollections.

- **Flatten** combines PCollections of the same type into a single PCollection.

- **Windowing and Triggers** are used to define mechanisms for window operations.

To illustrate some of these features, we consider an environmental sensor example in which each event consists of an instrument type, location, and numerical reading. We compute the average temperature for sensors of type `tempsensor` for each location using a sliding window. For the sake of illustration, we use an imaginary pub-sub system to get the events from the instrument stream. We suppose that the events are delivered to our system in the form of a Java object from the class `InstEvnt` declared in Beam as follows.

```
@DefaultCoder(AvroCoder.class)
static class InstEvent{
        @Nullable String instType;
        @Nullable String location;
        @Nullable Double reading;
        public InstEvent( ....)
        public String getInstType(){ ...}
        public String getLocation(){ ...}
        public String getReading(){ ...}
}
```

This class definition illustrates how a custom serializable type looks like in Beam. We can now create our stream from our fictitious pub-sub system with the following lines.

```
PCollection<InstEvent> input =
    pipeline.apply(PubsubIO.Read
        .timestampLabel(PUBSUB_TIMESTAMP_LABEL_KEY)
        .subscription(options.getPubsubSubscription()));
```

We next filter out all but the "tempsensor" events. While we are at it, let's convert the stream so that the output is a stream of key-value pairs corresponding to (location, reading). To do that, we need a special function to feed to the `ParDo` operator, as follows.

```
static class FilterAndConvert extends DoFn<InstEvent,
                                           KV<String, Double>> {
    @Override
    public void processElement(ProcessContext c) {
        InstEvent ev = c.element();
        if (ev.getInstType() == "tempsensor")
            c.output(KV<String, Double>.
                        of(ev.getLocation(), ev.getReading()));
    }
}
```

Now we can apply a `FilterAndConvert` operator to our input stream. We also create a sliding window of events of duration five minutes, created every two minutes. Note that the window is measured in terms of the timestamps on the events, rather than the processing time.

```
PCollection<KV<String, Float>> reslt =
    input.apply(Pardo.of(new FilterAndConvert())
        .apply(Window.<KV<String, Double>> into(SlidingWindows.of(
                Duration.standardMinutes(5))
                .every(Duration.standardMinutes(2))))
```

Our stream `reslt` is now a `KV<String,Double>` type, and we can apply a GroupByKey and Combine operation to reduce this to a `KV<String,Double>`, with each location key mapping to the average temperature. Beam provides several variations of this simple MapReduce operation. One is perfect for this case: `Mean.perKey()`, which combines both steps in a single transformation.

```
PCollection<KV<String, Double>> avetemps
    = reslt.apply(Mean.<String, Double>perKey());
```

We can now take the set of average temperatures for each window and send them to an output file.

```
PCollection<String> outstrings =
    avetemps.apply(Pardo.of(new KVToString())
            .apply(TextIO.Write.named("WritingToText")
            .to("/my/path/to/temps")
            .withSuffix(".txt"));
```

We must define the function class `KVToString()` in a manner similar to the FilterAndConvert class above. We call attention to two points. First, we have used an implicit trigger that generates the means and output at the end of the window. Second, because the windows overlap, events end up in more than one window.

Beam has several other types of triggers. For example, you can have a data-driven trigger that looks at the data as it is coming and fires when some condition you have set is met. Another type is based on a concept introduced by Google Dataflow called the **watermark**. A watermark is based on event time and is used to emit results when the system estimates that it has seen all the data in a given window. You can use a variety of methods to define triggers, based on different ways of specifying the watermark. We refer you to the Google Dataflow documents for details `cloud.google.com/dataflow/`.

9.8   Apache Flink

Finally we describe the open source Apache Flink stream processing framework. Flink can be used in its own right; in addition, a component called the Apache Flink Runner can be used to execute Beam pipelines. (Many of the same core concepts exist in Flink and Beam.)

As with the other systems described in this chapter, Flink takes input streams from one or more sources, which it connects by a directed graph to a set of sinks, and runs on the Java virtual machine. It supports both Java and Scala APIs. (There is also an incomplete **Flink Python API**, with similarities to Spark Streaming.) To illustrate the use of this API, we show a Flink implementation of our instrument filter from the Beam example. The Flink Kinesis Producer is still a work in progress, so we tested this code by reading a stream from a CSV file. Since the Flink data types do not include the Python dictionary/JSON types, we use here a simple tuple format. Each line in the input stream has this structure:

```
instrument-type string, location string, the word "value", float
```

For example:

```
tempsensor, pine street and second, value, 72.3
```

After reading from the file (or Kinesis shard), the records in the stream data are now four-tuples of type (STRING, STRING, STRING, FLOAT). The core of the Flink version of the temperature sensor averager is as follows.

```
class MeanReducer(ReduceFunction):
    def reduce(self, x, y):
        return (x[0], x[1], x[2], x[3] + y[3], x[4] + y[4])

env = get_environment()
data = env.add_source(FlinkKinesisProducer( ? ) ? )

results = data \
    .filter(lambda x: x[0]=='tempsensor') \
    .map(lambda x: (x[0], x[1], x[2], x[3], 1.0)) \
    .group_by(1) \
    .reduce(MeanReducer()) \
    .map(lambda x: 'location: '+x[1]+' average temp%f' %(x[3]/x[4]))
```

The filter operation is identical to the Spark Streaming case. After filtering the data, we turn each record into a five-tuple by appending 1.0 to the end of the four-tuple. The group_by and reduce calls use the MeanReducer function. The group_by is a signal to shuffle these so that they are keyed by field in position 1, which corresponds to the location string. We then apply the reduction to each of the grouped tuple sets. This operation is the same as the reduceByKey function in the Spark Streaming example. The final map converts each element to a string that gives the average temperature for each location.

Not shown in this example are Flink's windowing operators (similar to Beam's) and its underlying execution architecture. In a manner similar to the other systems described here, Flink parallelizes the stream and tasks during execution. For example, we can view our temperature sensor example as a set of tasks that may be executed in parallel, as shown in figure 9.10 on the following page.

The Flink distributed execution engine is based on a standard master worker model. A Flink source program is compiled into an execution data flow graph and sent to a job manager node by a client system. The job manager executes the stream and transformations on remote Java virtual machines that run a task manager. The task manager partitions its available resources into task slots, to which the individual tasks defined by the graph execution nodes are assigned. The job manager and task managers manage the data communication streams between the graph nodes. The Apache Flink documentation [2] explains this execution model, Flink windowing, and other details of the programming model.

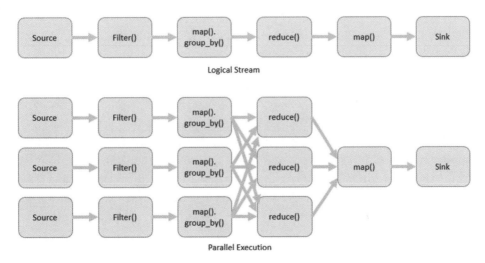

Figure 9.10: Flink logical task view and parallel execution view.

9.9   Summary

We have examined and illustrated the use of five different systems: Spark Streaming with Kinesis, Azure Stream Analytics, Storm/Heron, Google Dataflow/Beam, and Flink. Each has been used in critical production deployments and proven successful for its intended applications. All share some of the same concepts and create pipelines in similar ways. There are also differences: for example, Storm/Heron explicitly constructs graphs from nodes and edges, whereas the others use a functional style of pipeline composition.

Conceptually the greatest difference arises when comparing Spark Streaming with the others and, in particular, with Beam. Akidau and Perry make a compelling argument [59] for the superiority of the Beam model over Spark Streaming. They point out that Spark is a batch system for which a streaming mode has been attached, whereas Beam was designed from the ground up to be streaming with obvious batch capabilities. Spark windowing is based on the RDD in the DStream, which is clearly not as flexible as Beam windows. A more significant point concerns Beam's recognition that event time and processing time are not the same. This difference becomes critical in dealing with out-of-order events, which clearly can arise in widely distributed situations. Beam's introduction of event-time windows, triggers, and watermarks are a major contribution that clarifies important correctness issues when events are out of order, while still allowing the generation of approximate results in a timely manner.

The open source streaming tools that we have described here do not cover all of the scientific cases that we mentioned early in the chapter. For example, the ADIOS application requires the processing of events that are each of a size well beyond the limits of Spark Streaming, Storm/Heron, and Beam. In other scientific applications, the volume and rate of data generation are so large that we cannot keep data at rest for long. For example, the data to be produced by the Square Kilometer Array will be far too large to contemplate keeping [211]; therefore, the data will have to be processed immediately to produce a reduced stream. Another important aspect of large-scale streaming scientific data analysis is computational steering: the need for a human or smart processes to analyze the data stream for quality or relevance and then make rapid adjustments to the source instruments or simulations. Such requirements place further demands on streaming systems.

Science does not have the deep pockets of Google or Amazon when it comes to IT resources. Budgets are dominated by massive experimental facilities and supercomputers, and projects tend to produce custom designed solutions. And each experimental domain is sufficiently unique that few common tools beyond MPI exist. This is a world of bespoke software systems.

One can argue that the Twitter, Google, and Apache projects discussed here are also custom built for the problems that each is designed to solve. But each project has many project committers, often supported by companies to ensure that the software works for different problems. This situation does not mean that the cloud and open source tools are of no value to the world of "big science." The Google and Amazon clouds have both been used to process data from the Large Hadron Collider. We expect many of the cloud streaming technologies described here to eventually be adapted to scientific solutions. Likewise, we expect tools developed in scientific laboratories to migrate to the cloud.

9.10 Resources

There are excellent online tutorials on Amazon Kinesis `aws.amazon.com/kinesis` and Azure Stream Analytics `docs.microsoft.com/en-us/azure/stream-analytics`. Amazon's examples include demonstrations of log analysis, mobile data capture and game data feeds. The Azure demos include examples of how to do sentiment analysis of Twitter streams as well as fraud analysis. Google Dataflow streaming also has good examples [23], including Wikipedia session analysis, traffic routes, and maximum flows from traffic sensors.

Chapter 10

Machine Learning in the Cloud

> *"Learning is any change in a system that produces a more or less permanent change in its capacity for adapting to its environment."*
>
> —Herbert Simon, *The Sciences of the Artificial*

Machine learning has become central to applications of cloud computing. While machine learning is considered part of the field of artificial intelligence, it has roots in statistics and mathematical optimization theory and practice. In recent years it has grown in importance as a number of critical application breakthroughs have taken place. These include human-quality speech recognition [144] and real-time automatic language translation [95], computer vision accurate and fast enough to propel self-driving cars [74], and applications of reinforcement learning that allow machines to master some of the most complex human games, such as Go [234].

What has enabled these breakthroughs has been a convergence of the availability of big data plus algorithmic advances and faster computers that have made it possible to train even deep neural networks. The same technology is now being applied to scientific problems as diverse as predicting protein structure [180], predicting the pharmacological properties of drugs [60], and identifying new materials with desired properties [264].

In this chapter we introduce some of the major machine learning tools that are available in public clouds, as well as toolkits that you can install on a private cloud. We begin with our old friend Spark and its machine learning (ML) package, and then move to Azure ML. We progress from the core "classical" ML tools, including logistic regression, clustering, and random forests, to take a brief look at deep learning and deep learning toolkits. Given our emphasis on Python, the reader may expect us to cover the excellent Python library *scikit-learn*. However, *scikit-learn* is well covered elsewhere [253], and we introduced several of its ML methods in our microservice-based science document classifier example in chapter 7. We describe the same example, but using different technology, later in this chapter.

10.1 Spark Machine Learning Library (MLlib)

Spark MLlib [198], sometimes referred to as Spark ML, provides a set of high-level APIs for creating ML pipelines. It implements four basic concepts.

- **DataFrames** are containers created from Spark RDDs to hold vectors and other structured types in a manner that permits efficient execution [45]. Spark DataFrames are similar to Pandas DataFrames and share some operations. They are distributed objects that are part of the execution graph. You can convert them to Pandas DataFrames to access them in Python.

- **Transformers** are operators that convert one DataFrame to another. Since they are nodes on the execution graph, they are not evaluated until the entire graph is executed.

- **Estimators** encapsulate ML and other algorithms. As we describe in the following, you can use the `fit(...)` method to pass a DataFrame and parameters to a learning algorithm to create a model. The model is now represented as a Transformer.

- A **Pipeline** (usually linear, but can be a directed acyclic graph) links Transformers and Estimators to specify an ML workflow. Pipelines inherit the `fit(...)` method from the contained estimator. Once the estimator is trained, the pipeline is a model and has a `transform(...)` method that can be used to push new cases through the pipeline to make predictions.

Many transformers exist, for example to turn text documents into vectors of real numbers, convert columns of a DataFrame from one form to another, or split DataFrames into subsets. There are also various kinds of estimators, ranging from those that transform vectors by projecting them onto principal component vectors, to n-gram generators that take text documents and return strings of n consecutive words. Classification models include logistic regression, decision tree classifiers, random forests, and naive Bayes. The family of clustering methods includes k-means and latent Dirichlet allocation (LDA). The MLlib online documentation provides much useful material on these and related topics [29] .

10.1.1 Logistic Regression

The example that follows employs a method called **logistic regression** [103], which we introduce here. Suppose we have a set of feature vectors $x_i \in R^n$ for i in $[0, m]$. Associated with each feature vector is a binary outcome y_i. We are interested in the conditional probability $P(y = 1|x)$, which we approximate by a function $p(x)$. Because $p(x)$ is between 0 and 1, it is not expressible as a linear function of x, and thus we cannot use regular linear regression. Instead, we look at the "odds" expression $p(x)/(1 - p(x))$ and guess that its log is linear. That is:

$$ln\left(\frac{p(x)}{1 - p(x)}\right) = b_0 + b \cdot x,$$

where the offset b_0 and the vector $b = [b_1, b_2, ...b_n]$ define a hyperplane for linear regression. Solving this expression for $p(x)$ we obtain:

$$p(x) = \frac{1}{1 + e^{-(b_0 + b \cdot x)}}$$

We then predict $y = 1$ if $p(x) > 0$ and zero otherwise. Unfortunately, finding the best b_0 and b is not as easy as in the case of linear regression. However, simple Newton-like iterations converge to good solutions if we have a sample of the feature vectors and known outcomes.

(We note that the **logistic function** $\sigma(t)$ is defined as follows:

$$\sigma(t) = \frac{e^t}{e^t + 1} = \frac{1}{1 + e^{-t}}$$

It is used frequently in machine learning to map a real number into a probability range $[0, 1]$; we use it for this purpose later in this chapter.)

10.1.2 Chicago Restaurant Example

To illustrate the use of Spark MLlib, we apply it to an example from the Azure HDInsight tutorial [195], namely predicting whether restaurants pass or fail health inspections based on the free text of an inspector's comments. We provide two versions of this example in notebook 18: the HDInsight version and a version that runs on any generic Spark deployment. We present the second here.

The data, from the City of Chicago Data Portal `data.cityofchicago.org`, are a set of restaurant heath inspection reports. Each inspection report contains a report number, the name of the owner of the establishment, the name of the establishment, the address, and an outcome ("Pass," "Fail," or some alternative such as "out of business" or "not available"). It also contains the (free-text) English comments from the inspector.

We first read the data. If we are using Azure HDInsight, we can load it from blob storage as follows. We use a simple function `csvParse` that takes each line in the CSV file and parses it using Python's `csv.reader()` function.

```
inspections = spark.sparkContext.textFile( \
    'wasb:///HdiSamples/HdiSamples/FoodInspectionData/
    Food_Inspections1.csv').map(csvParse)
```

The version of the program in notebook 18 uses a slightly reduced dataset. We have eliminated the address fields and some other data that we do not use here.

```
inspections = spark.sparkContext.textFile(
    '/path-to-reduced-data/Food_Inspections1.csv').map(csvParse)
```

We want to create a **training set** from a set of inspection reports that contain outcomes, for use in fitting our logistic regression model. We first convert the RDD containing the data, `inspections`, to create a DataFrame, `df`, with four fields: record id, restaurant name, inspection result, and any recorded violations.

```
schema = StructType([StructField("id", IntegerType(), False),
                     StructField("name", StringType(), False),
                     StructField("results", StringType(), False),
                     StructField("violations", StringType(), True)])

df = spark.createDataFrame(inspections.map(\
        lambda l: (int(l[0]), l[2], l[3], l[4])) , schema)
df.registerTempTable('CountResults')
```

If we want to look at the first few elements, we can apply the **show()** function to return values to the Python environment.

```
df.show(5)
+-------+-------------------+----------------+-------------------+
|     id|               name|         results|         violations|
+-------+-------------------+----------------+-------------------+
|1978294|KENTUCKY FRIED CH...|           Pass|32. FOOD AND NON-...|
|1978279|         SOLO FOODS|Out of Business|                   |
|1978275|SHARKS FISH & CHI...|           Pass|34. FLOORS: CONST...|
|1978268|CARNITAS Y SUPERM...|           Pass|33. FOOD AND NON-...|
|1978261|           WINGSTOP|           Pass|                   |
+-------+-------------------+----------------+-------------------+
only showing top 5 rows
```

Fortunately for the people of Chicago, it seems that the majority of the inspections result in passing grades. We can use some DataFrame operations to count the passing and failing grades.

```
print("Passing = %d"%df[df.results == 'Pass'].count())
print("Failing = %d"%df[df.results == 'Fail'].count())

Passing = 61204
Failing = 20225
```

To train a logistic regression model, we need a DataFrame with a binary label and feature vector for each record. We do not want to use records associated with "out of business" or other special cases, so we map "Pass" and "Pass with conditions" to 1, "Fail" to 0, and all others to -1, which we filter out.

```
def labelForResults(s):
    if s == 'Fail':
        return 0.0
    elif s == 'Pass w/ Conditions' or s == 'Pass':
        return 1.0
    else:
        return -1.0

label = UserDefinedFunction(labelForResults, DoubleType())
labeledData = df.select(label(df.results).alias('label'), \
                        df.violations).where('label >= 0')
```

We now have a DataFrame with two columns, `label` and `violations` and we are ready to create and run the Spark MLlib pipeline that we will use to train our logistic regression model, which we do with the following code.

```
# 1) Define pipeline components
#    a) Tokenize 'violations' and place result in new column 'words'
tokenizer = Tokenizer(inputCol="violations", outputCol="words")
#    b) Hash 'words' to create new column of 'features'
hashingTF = HashingTF(inputCol="words" , outputCol="features")
#    c) Create instance of logistic regression
lr = LogisticRegression(maxIter=10, regParam=0.01)

# 2) Construct pipeline: tokenize, hash, logistic regression
pipeline = Pipeline(stages=[tokenizer, hashingTF, lr])

# 3) Run pipeline to create model
model = pipeline.fit(labeledData)
```

We first (1) define our three pipeline components, which (a) tokenize each `violations` entry (a text string) by reducing it to lower case and splitting it into

195

a vector of words; (b) convert each word vector into a vector in R^n for some n, by applying a hash function to map each word token into a real number value (the new vectors have length equal to the size of the vocabulary, and are stored as sparse vectors); and (c) create an instance of logistic regression. We then (2) put everything into a pipeline and (3) fit the model with our labeled data.

Recall that Spark implements a graph execution model. Here, the pipeline created by the Python program is the graph; this graph is passed to the Spark execution engine by calling the `fit(...)` method on the pipeline. Notice that the `tokenizer` component adds a column `words` to our working DataFrame, and `hashingTF` adds a column `features`; thus, the working DataFrame has columns `ID, name, results, label, violations, words, features` when logistic regression is run. The names are important, as logistic regression looks for columns `label, features`, which it uses for training to build the model. The trainer is iterative; we give it 10 iterations and an algorithm-dependent value of `0.01`.

We can now test the model with a separate test collection as follows.

```
testData = spark.sparkContext.textFile(
            '/data_path/Food_Inspections2.csv')\
        .map(csvParse) \
        .map(lambda l: (int(l[0]), l[2], l[3], l[4]))
testDf = spark.createDataFrame(testData, schema).
    where("results = 'Fail' OR results = 'Pass' OR \
            results = 'Pass w/ Conditions'")
predictionsDf = model.transform(testDf)
```

The logistic regression model has appended several new columns to the data frame, including one called `prediction`. To test our prediction success rate, we compare the `prediction` column with the `results` column.

```
numSuccesses = predictionsDf.where(\
            """(prediction = 0 AND results = 'Fail') OR \
            (prediction = 1 AND (results = 'Pass' OR \
            results = 'Pass w/ Conditions'))""").count()
numInspections = predictionsDf.count()
print("There were %d inspections and there were %d predictions"\
    %(numInspections,numSuccesses))
print("This is a %2.2f sucess rate"\
    %(float(numSuccesses) / float(numInspections) * 100))
```

We see the following output:

```
There were 30694 inspections and there were 27774 predictions
This is a 90.49\% success rate
```

Before getting too excited about this result, we examine other measures of success, such as **precision** and **recall**, that are widely used in ML research. When applied to our ability to predict failure, recall is the probability that we predicted as failing a randomly selected inspection from those with failing grades. As detailed in notebook 18, we find that our recall probability is only 67%. Our ability to predict failure is thus well below our ability to predict passing. The reason may be that other factors involved with failure are not reflected in the report.

10.2 Azure Machine Learning Workspace

Azure Machine Learning is a cloud portal for designing and training machine learning cloud services. It is based on a drag-and-drop component composition model, in which you build a solution to a machine learning problem by dragging parts of the solution from a pallet of tools and connecting them together into a workflow graph. You then train the solution with your data. When you are satisfied with the results, you can ask Azure to convert your graph into a running web service using the model you trained. In this sense Azure ML provides customized machine learning as an on-demand service. This is another example of serverless computation. It does not require you to deploy and manage your own VMs; the infrastructure is deployed as you need it. If your web service needs to scale up because of demand, Azure scales the underlying resources automatically.

To illustrate how Azure ML works, we return to an example that we first considered in chapter 7. Our goal is to train a system to classify scientific papers, based on their abstracts, into one of five categories: physics, math, computer science, biology, or finance. As training data we take a relatively small sample of abstracts from the arXiv online library `arxiv.org`. Each sample consists of a triple: a classification from arXiv, the paper title, and the abstract. For example, the following is the record for a 2015 paper in physics [83].

```
[ 'Physics',
'A Fast Direct Sampling Algorithm for Equilateral Closed Polygons. (arXiv:1510.02466v1 [cond-
mat.stat-mech])',
'Sampling equilateral closed polygons is of interest in the statistical study of ring polymers. Over
the past 30 years, previous authors have proposed a variety of simple Markov chain algorithms
(but have not been able to show that they converge to the correct probability distribution) and
complicated direct samplers (which require extended-precision arithmetic to evaluate numerically
unstable polynomials). We present a simple direct sampler which is fast and numerically stable.'
]
```

This example also illustrates one of the challenges of the classification problem: science has become wonderfully multidisciplinary. The topic given for this sample paper in arXiv is "condensed matter," a subject in physics. Of the four authors, however, two are in mathematics institutes and two are from physics departments, and the abstract refers to algorithms that are typically part of computer science. A human reader might reasonably consider the abstract to be describing a topic in mathematics or computer science. (In fact, multidisciplinary physics papers were so numerous in our dataset that we removed them in the experiment below.)

Let us start with a solution in Azure ML based on a multiclass version of the logistic regression algorithm. Figure 10.1 shows the graph of tasks. To understand this workflow, start at the top, which is where the data source comes into the picture. Here we take the data from Azure blob storage, where we have placed a large subset of our arXiv samples in a CSV file. Clicking the $\boxed{\texttt{Import Data}}$ box opens the window that allows us to identify the URL for the input file.

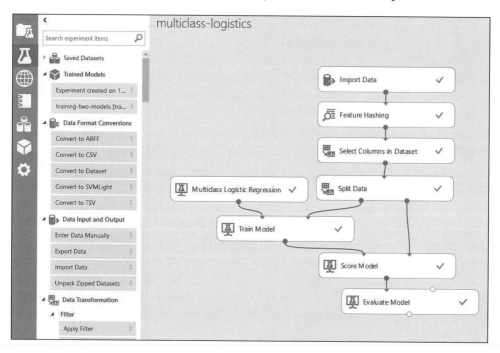

Figure 10.1: Azure ML graph used to train a multiclass logistic regression model.

The second box down, `Feature Hashing`, builds a vectorizer based on the vocabulary in the document collection. This version comes from the Vowpal Wabbit library. Its role is to convert each document into a numerical vector corresponding to the key words and phrases in the document collection. This numeric representation is essential for the actual ML phase. To create the vector, we tell the feature hasher to look only at the abstract text. What happens in the output is that the vector of numeric values for the abstract text is appended to the tuple for each document. Our tuple now has a large number of columns: class, title, abstract, and `vector[0]`, ..., `vector[n-1]`, where n is the number of features. To configure the algorithm, we select two parameters, a hashing bin size and an n-gram length.

Before sending the example to ML training, we remove the English text of the abstract and the title, leaving only the class and the vector for each document. We accomplish this with a `Select Columns in Dataset`. Next we split the data into two subsets: a training subset and a test subset. (We specify that `Split Data` should use 75% of the data for training and the rest for testing.)

Azure ML provides a good number of the standard ML modules. Each such module has various parameters that can be selected to tune the method. For all the experiments described here, we just used the default parameter settings. The `Train Model` component accepts as one input a binding to an ML method (recall this is not a dataflow graph); the other input is the projected training data. The output of the Train Model task is not data per se but a trained model that may also be saved for later use. We can now use this trained model to classify our test data. To this end, we use the `Score Model` component, which appends another new column to our table, Scored Label, providing the classification predicted by the trained model for each row.

To see how well we did, we use the `Evaluate Model` component, which computes a confusion matrix. Each row of the matrix tells us how the documents in that class were classified. Table 10.1 shows the confusion matrix for this experiment. Observe, for example, that a fair number of biology papers are classified as math. We attribute this to the fact that most biology papers in the archive are related to quantitative methods, and thus contain a fair amount of mathematics. To access the confusion matrix, or for that matter the output of any stage in the graph, click on the output port (the small circle) on the corresponding box to access a menu. Selecting *visualize* in that menu brings up useful information.

Table 10.1: Confusion matrix with only math, computer science, biology, and finance.

	bio	compsci	finance	math
bio	51.3	19.9	4.74	24.1
compsci	10.5	57.7	4.32	27.5
finance	6.45	17.2	50.4	25.8
math	6.45l	16.0	5.5	72

Now that we have trained the model, we can click the [Set Up Web Service] button (not visible, but at the bottom of the page) to turn the model into a web service. The Azure ML portal rearranges the graph by eliminating the split-train-test parts and leaves just the feature hashing, column selection, and the scoring based on the trained model. Two new nodes have been added: a web service input and a web service output. The result, with one exception, is shown in figure 10.2. The exception is that we have added a new **Select Columns** node so that we can remove the vectorized document columns from the output of the web service. We retain the original class, the predicted class, and the probabilities computed for the document being in a class.

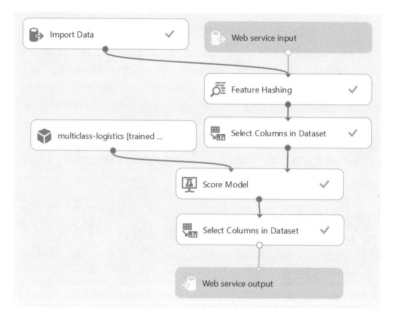

Figure 10.2: Web service graph generated by Azure ML, with an additional node to remove the vectorized document.

You can now try additional ML classifier algorithms simply by replacing the box *Multiclass Logistic Regression* with, for example, *Multiclass Neural Network* or *Random forest classifier*. Or, you can incorporate all three methods into a single web service that uses a majority vote ("consensus") method to pick the best classification for each document. As shown in figure 10.3, the construction of this consensus method is straightforward: we simply edit the web service graph for the multiclass logistic regression to add the trained models for the other two methods and then call a Python script to tie the three results together.

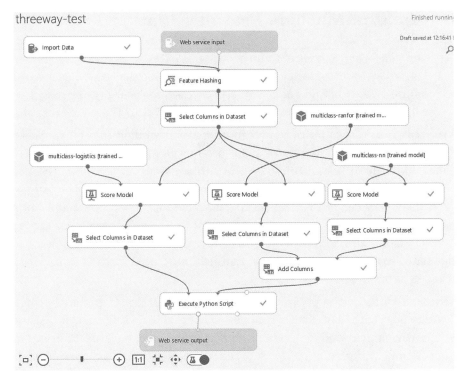

Figure 10.3: Modified web service graph based on a consensus model, showing three models and a Python script component, used to determine the consensus.

The Python script can simply compare the outputs from the three classifiers. If any two agree, then it selects that classification as a first choice and the classification that does not agree as a second choice. The results for the first choice, shown in table 10.2, are only modestly better than in the logistic regression case, but if we consider both the first and second choices, we reach 65% for biology, 72% for computer science, 60% for finance, and 88% for math. Notebook 19 contains this Python script as well as the code used to test and invoke the services and to compute the confusion matrices.

Table 10.2: Confusion matrix for the three-way classifier.

	bio	compsci	finance	math
bio	50.3	20.9	0.94	27.8
compsci	4.9	62.7	1.54	30.9
finance	5.6	9.9	47.8	36.6
math	3.91	13.5	2.39	80.3

10.3 Amazon Machine Learning Platform

The Amazon platform provides an impressive array of ML services. Each is designed to allow developers to integrate cloud ML into mobile and other applications. Three of the four are based on the remarkable progress that has been enabled by the deep learning techniques that we discuss in more detail in the next section.

Amazon Lex allows users to incorporate voice input into applications. This service is an extension of Amazon's Echo product, a small networked device with a speaker and a microphone to which you can pose questions about the weather and make requests to schedule events, play music, and report on the latest news. With Lex as a service, you can build specialized tools that allow a specific voice command to Echo to launch an Amazon lambda function to execute an application in the cloud. For example, NASA has built a replica of the NASA Mars rover that can be controlled by voice commands, and has integrated Echo into several other applications around their labs [189].

Amazon Polly is the opposite of Lex: it turns text into speech. It can speak in 27 languages with a variety of voices. Using the Speech Synthesis Markup Language, you can carefully control pronunciation and other aspects of intonation. Together with Lex, Polly makes a first step toward conversational computing. Polly and Lex do not do real-time, voice-to-voice language translation the way Skype does, but together they provide a great platform to deliver such a service.

Amazon Rekognition is at the cutting edge of deep learning applications. It takes an image as input and returns a textual description of the items that it sees in that image. For example, given an image of a scene with people, cars, bicycles, and animals, Rekognition returns a list of those items, with a measure of certainty associated with each. The service is trained with many thousands of captioned images in a manner not unlike the way natural language translation systems are trained: it considers a million images containing a cat, each with an associated caption that mentions "cat," and a model association is formed. Rekognition can also perform detailed facial analysis and comparisons.

The **Amazon Machine Learning** service, like Azure ML, can be used to create a predictive model based on training data that you provide. However, it requires much less understanding of ML concepts than does Azure ML. The Amazon Machine Learning dashboard presents the list of experiments, models, and data sources from your previous Amazon Machine Learning work. From the dashboard you can define data sources and ML models, create evaluations, and run batch predictions.

Using Amazon Machine Learning is easy. For example, we used it to build a predictive model from our collection of scientific articles in under an hour. One reason that it is so easy to use is that the options are simple. You can build only three types of models—regression, binary classification, or multiclass classification—and in each case, Amazon Machine Learning provides a single model. In the case of multiclass classification, it is multinomial logistic regression with a stochastic gradient descent optimizer. And it works well. Using the same test and training data as earlier, we obtained the results shown in table 10.3. Although the trained Amazon Machine Learning classifier failed to recognize any computational finance papers, it beat our other classifiers in the other categories. Amazon Labs has additional excellent examples [44].

Table 10.3: Confusion matrix for the science document classifier using Amazon ML.

	bio	compsci	finance	math
bio	62.0	19.9	0.0	18.0
compsci	3.8	78.6	0.0	17.8
finance	6.8	2.5	0.0	6.7
math	3.5	11.9	0.0	84.6

Amazon Machine Learning is also fully accessible from the Amazon REST interface. For example, you can create a ML model using Python as follows.

```
response = client.create_ml_model(
    MLModelId='string',
    MLModelName='string',
    MLModelType='REGRESSION'|'BINARY'|'MULTICLASS',
    Parameters={
        'string': 'string'
    },
    TrainingDataSourceId='string',
    Recipe='string',
    RecipeUri='string'
)
```

The parameter ModelID is a required, user-supplied, unique identifier; other parameters specify, for example, the maximum allowed size of the model, the maximum number of passes over the data in building the model, and a flag to tell the learners to shuffle the data. The training data source identifier is a data recipe or URI for a recipe in S3. A recipe is a JSON-like document that describes how to transform the datasets for input while building the model. (Consult the Amazon Machine Learning documents for more details.) For our science document example, we used the default recipe generated by the portal.

10.4 Deep Learning: A Shallow Introduction

The use of **artificial neural networks** for machine learning tasks has been common for at least 40 years. Mathematically, a neural network is a method of approximating a function. For example, consider the function that takes an image of a car as input and produces the name of a manufacturer as output. Or, consider the function that takes the text of a scientific abstract and outputs the most likely scientific discipline to which it belongs. In order to be a computational entity, our function and its approximation need a numerical representation. For example, suppose our function takes as input a vector of three real numbers and returns a vector of length two. Figure 10.4 is a diagram of a neural net with one hidden layer representing such a function.

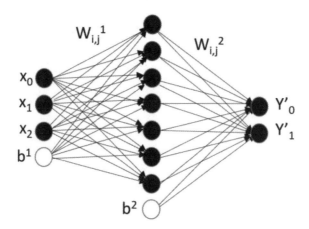

Figure 10.4: Neural network with three inputs, one hidden layer, and two outputs.

In this schematic representation, the lines represent numerical weights connecting the inputs to the n interior neurons, and the terms b are offsets. Mathematically the function is given by the following equations.

$$a_j = f\left(\sum_{i=0}^{2} x_i W_{i,j}^1 + b_j\right) \qquad for \quad j = 1, n$$

$$y_j' = f'\left(\sum_{i=0}^{n} a_i W_{i,j}^2 + b_j^2\right) \qquad for \quad j = 0, 1$$

The functions f and f' are called the **activation functions** for the neurons. Two commonly used activation functions are the logistic function $\sigma(t)$ that we introduced at the beginning of this chapter and the rectified linear function:

$$\text{relu}(x) = \max(0, x).$$

Another common case is the hyperbolic tangent function

$$\tanh(x) = \frac{e^x - e^{-x}}{e^x + e^{-x}}$$

An advantage of $\sigma(x)$ and $\tanh(x)$ is that they map values in the range $(-\infty, \infty)$ to the range $(0, 1)$, which corresponds to the idea that a neuron is either on or off (not-fired or fired). When the function represents the probability that an input corresponds to one of the outputs, we use a version of the logistic function that ensures that the probabilities all sum to one.

$$\text{softmax}(x)_j = \frac{1}{1 + \sum_{k \neq j} e^{x_k - x_j}}$$

This formulation is commonly used in multiclass classification, including in several of the examples we have studied earlier.

The trick to making the neural net truly approximate our desired function is picking the right values for the weights. There is no closed form solution for finding the best weights, but if we have a large number of labeled examples (x^i, y^i), we can try to minimize the cost function.

$$C(x^i, y^i) = \sum \|y^i - y'(x^i)\|$$

The standard approach is to use a variation of gradient descent, specifically **back propagation**. We do not provide details on this algorithm here but instead refer you to two outstanding mathematical treatments of deep learning [143, 210].

10.4.1 Deep Networks

An interesting property of neural networks is that we can stack them in layers as illustrated in figure 10.5 on the next page. Furthermore, using the deep learning toolkits that we discuss in the remainder of this chapter, we can construct such networks with just a few lines of code. In this chapter, we introduce three deep learning toolkits. We first illustrate how each can be used to define some of the standard deep networks and then, in later sections, describe how to deploy and apply them in the cloud.

MXNet `github.com/dmlc/mxnet` is the first deep learning toolkit that we consider. Using MXNet, the network in figure 10.5 would look as follows.

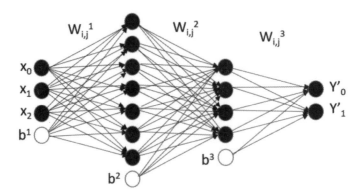

Figure 10.5: Neural network with three inputs, two hidden layers, and two outputs.

```
data = mx.symbol.Variable('x')
layr1= mx.symbol.FullyConnected(data=data,name='W1',num_hidden=7)
act1 = mx.symbol.Activation(data=layr1,name='relu1',act_type="relu")
layr2= mx.symbol.FullyConnected(data=act1,name='W2',num_hidden=4)
act2 = mx.symbol.Activation(data=layr2,name='relu2',act_type="relu")
layr3= mx.symbol.FullyConnected(data=act2, name='W3',num_hidden=2)
Y    = mx.symbol.SoftmaxOutput(data = layr3,name='softmax')
```

The code creates a stack of fully connected networks and activations that exactly describe our diagram. In the following section we return to the code needed to train and test this network.

The term **deep neural network** generally refers to networks with many layers. Several special case networks also have proved to be of great value for certain types of input data.

10.4.2 Convolutional Neural Networks

Data with a regular spatial geometry such as images or one-dimensional streams are often analyzed with a special class of network called a **convolutional neural network** or CNN. To explain CNNs, we use our second example toolkit, **TensorFlow** `tensorflow.org`, which was open sourced by Google in 2016. We consider a classic example that appears in many tutorials and is well covered in that provided with TensorFlow `tensorflow.org/tutorials`.

Suppose you have thousands of 28×28 black and white images of handwritten digits and you want to build a system that can identify each. Images are strings of bits, but they also have a lot of local two-dimensional structure such as edges and holes. In order to find these patterns, we examine each of the many 5×5 windows in each image individually. To do so, we train the system to build a 5×5 template array $W1$ and a scalar offset b that together can be used to reduce each 5×5 window to a point in a new array conv by the following formula.

$$\text{conv}_{p,q} = \sum_{i,k=-2}^{2} W_{i,k} \text{image}_{p-i,q-k} + b$$

(The image is padded near the boundary points in the formula above so that none of the indices are out of bounds.) We next modify the conv array by applying the relu function to each x in the conv array so that it has no negative values. The final step, *max pooling*, simply computes the maximum value in each 2×2 block and assigns it to a smaller 14×14 array. The most interesting part of the convolutional network is that we do not use one 5×5 $W1$ template but 32 of them in parallel, producing 32 14×14 results, pool1, as illustrated in figure 10.6.

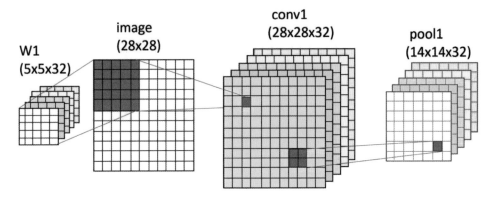

Figure 10.6: Schematic of how a convolutional neural net processes an image.

When the network is fully trained, each of the 32 5×5 templates in $W1$ is somehow different, and each selects for a different set of features in the original image. One can think of the resulting stack of 32 14×14 arrays (called pool1) as a type of transform of the original image, which works much like a Fourier transform to separate a signal in space and time and transform it into frequency space. This is not what is going on here; but if you are familiar with these transforms, the analogy may be helpful.

We next apply a second convolutional layer to pool1, but this time we apply 64 sets of 5×5 filters to each of the 32 pool1 layers and sum the results to obtain 64 new 14×14 arrays. We then reduce these with max pooling to 64 7×7 arrays called `pool2`. From there we use a dense "all-to-all" layer and finally reduce it to 10 values, each representing the likelihood that the image corresponds to a digit 0 to 9. The TensorFlow tutorial defines two ways to build and train this network; figure 10.7 is from the community-contributed library called *layers*.

```
input_layer = tf.reshape(features, [-1, 28, 28, 1])

conv1 = tf.layers.conv2d(
     inputs=input_layer,
     filters=32,
     kernel_size=[5, 5],
     padding="same",
     activation=tf.nn.relu)

pool1 = tf.layers.max_pooling2d(inputs=conv1, \
                                pool_size=[2, 2], strides=2)

conv2 = tf.layers.conv2d(
     inputs=pool1,
     filters=64,
     kernel_size=[5, 5],
     padding="same",
     activation=tf.nn.relu)

pool2 = tf.layers.max_pooling2d(inputs=conv2, \
                                pool_size=[2, 2], strides=2)
pool2_flat = tf.reshape(pool2, [-1, 7 * 7 * 64])

dense = tf.layers.dense(inputs=pool2_flat, \
                        units=1024, activation=tf.nn.relu)

logits = tf.layers.dense(inputs=dense, units=10)
```

Figure 10.7: TensorFlow two convolutional layer digit recognition network

As you can see, these operators explicitly describe the features of our CNNs. The full program is in the TensorFlow examples tutorial layers directory in file `cnn_mnist.py`. If you would rather see a version of the same program using lower-level TensorFlow operators, you can find an excellent Jupyter notebook version in the Udacity deep learning course material [3]. CNNs have many applications in image analysis. One excellent science example is the solution to the Kaggle Galaxy Zoo Challenge, which asked participants to predict how Galaxy Zoo users would classify images of galaxies from the Sloan Digital Sky Survey. Dieleman [111] describes the solution, which uses four convolutional layers and three dense layers.

10.4.3 Recurrent Neural Networks.

Recurrent neural networks (RNNs) are widely used in language modeling problems, such as predicting the next word to be typed when texting or in automatic translation systems. RNNs can learn from sequences that have repeating patterns. For example, they can learn to "compose" text in the style of Shakespeare [168] or even music in the style of Bach [183]. They have also been used to study forest fire area coverage [94] and cycles of drought in California [178].

The input to the RNN is a word or signal, along with the state of the system based on words or signals seen so far; the output is a predicted list and a new state of the system, as shown in figure 10.8.

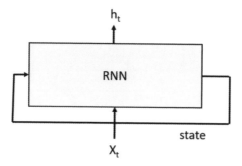

Figure 10.8: Basic RNN with input stream x and output stream h.

Many variations of the basic RNN exist. One challenge for RNNs is ensuring that the state tensors retain enough long-term memory of the sequence so that patterns are remembered. Several approaches have been used for this purpose. One popular method is the Long-Short Term Memory (LSTM) version that is defined by the following equations, where the input sequence is x, the output is h, and the state vector is the pair $[c, h]$.

$$i_t = \sigma(W^{(xi)}x_t + W^{(hi)}h_{t-1} + W^{(ci)}c_{t-1} + b^{(i)})$$
$$f_t = \sigma(W^{(xf)}x_t + W^{(hf)}h_{t-1} + W^{(cf)}c_{t-1} + b^{(f)})$$
$$c_t = f_t \cdot c_{t-1} + i_t \cdot \tanh(W^{(xc)}x_t + W^{(hc)}h_{t-1} + b^{(c)})$$
$$o_t = \sigma(W^{(xo)}x_t + W^{(ho)}h_{t-1} + W^{(co)}c_t + b^{(o)})$$
$$h_t = o_t \cdot \tanh(c_t)$$

Olah provides an excellent explanation of how RNNs work [213]. We adapt one of his illustrations to show in figure 10.9 on the next page how information flows in our network. Here we use the vector concatenation notation `concat` as follows to compose the various W matrices and thus obtain a more compact representation of the equations.

$$\sigma(\text{concat}(x, h, c)) = \sigma(W[x, h, c] + b) = \sigma(W^{(x)}x + W^{(h)}h + W^{(c)}c + b).$$

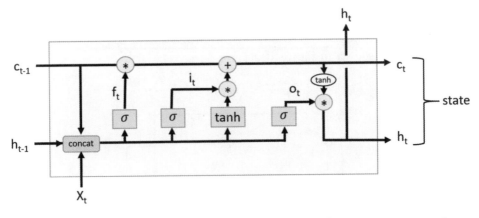

Figure 10.9: LSTM information flow, adapted from Olah [213] to fit equations in the text.

We use a third toolkit, the **Microsoft Cognitive Toolkit** (formerly known as the Computational Network Toolkit **CNTK**), to illustrate the application of RNNs. Specfically, we consider a *sequence-to-sequence* LSTM from the Microsoft Cognitive Toolkit distribution that is trained with input from financial news items such as "shorter maturities are considered a sign of rising rates because portfolio managers can capture higher rates sooner" and "j. p. <unk> vice chairman of grace and co. which holds an interest in this company was elected a director." Following training, this network can be used to generate other sentences with a similar structure.

To illustrate what this network can do, we saved the trained W and b arrays and two other arrays that together define its structure. We then loaded these arrays into the Python version of the RNN shown on the next page, which we created by transcribing the equations above.

```
def rnn(word, old_h, old_c):
    Xvec = getvec(word, E)
    i = Sigmoid(np.matmul(WXI, Xvec) +
                np.matmul(WHI, old_h) + WCI*old_c + bI)
    f = Sigmoid(np.matmul(WXF, Xvec) +
                np.matmul(WHF, old_h) + WCF*old_c + bF)
    c = f*old_c + i*(np.tanh(np.matmul(WXC, Xvec) +
                            np.matmul(WHC, old_h) + bC))
    o = Sigmoid(np.matmul(WXO, Xvec)+
                np.matmul(WHO, old_h)+ (WCO * c)+ bO)
    h = o * np.tanh(c)
    # Extract ordered list of five best possible next words
    q = h.copy()
    q.shape = (1, 200)
    output = np.matmul(q, W2)
    outlist = getwordsfromoutput(output)
    return h, c, outlist
```

As you can see, this code is almost a literal translation of the equations. The only difference is that the code has as input a text string for the input word, while the equations take a vector encoding of the word as input. The RNN training generated the encoding matrix E, which has the nice property that the ith column of the matrix corresponds to the word in the ith position in the vocabulary list. The function getvec(word, E) takes the embedding tensor E, looks up the position of the word in the vocabulary list, and returns the column vector of E that corresponds to that word. The output of one pass through the LSTM cell is the vector h. This is a compact representation of the words likely to follow the input text to this point. To convert this back into "vocabulary" space, we multiply it by another trained vector $W2$. The size of our vocabulary is 10,000, and the vector output is that length. The ith element of the output represents the relative likelihood that the ith word is the next word to follow the input so far. Our addition, Getwordsfromoutput, simply returns the top five candidate words, in order of likelihood.

To see whether this LSTM is truly a recurrent network, we provide the network with a starting word, let it suggest the next word, and repeat this process to construct a "sentence." In the code on the next page, we randomly pick one of the top three suggested by the network as the next word.

```
c = np.zeros(shape = (200, 1))
h = np.zeros(shape = (200, 1))
output = np.zeros(shape = (10000, 1))
word = 'my'
sentence= word
for _ in range(40):
    h, c, outlist = rnn(word, h, c)
    word = outlist[randint(0,3)]
    sentence = sentence + " " +word
print(sentence+".")
```

Testing this code with the start word "my" produced the following output.

```
my new rules which would create an interest position here unless
there should prove signs of such things too quickly although the
market could be done better toward paying further volatility where
it would pay cash around again if everybody can.
```

Using "the" as our start word produced the following.

```
the company reported third-quarter results reflecting a number
compared between N barrels including pretax operating loss
from a month following fiscal month ending july earlier
compared slightly higher while six-month cds increased
sharply tuesday after an after-tax loss reflecting a strong.
```

This RNN is hallucinating financial news. The sentences are obviously nonsense, but they are excellent examples of mimicry of the patterns that the network was trained with. The sentences end rather abruptly because of the 40-word limit in the code. If you let it go, it runs until the state vector for the sentence seems to break down. Try this yourself. To make it easy to play with this example, we have put the code in notebook 20 along with the 50 MB model data.

10.5 Amazon MXNet Virtual Machine Image

MXNet [92] `github.com/dmlc/mxnet` is an open source library for distributed parallel machine learning. It was originally developed at Carnegie Mellon, the University of Washington, and Stanford. MXNet can be programmed with Python, Julia, R, Go, Matlab, or C++ and runs on many different platforms, including clusters and GPUs. It is also now the deep learning framework of choice for Amazon [256]. Amazon has also released the **Amazon Deep Learning AMI** [13], which includes not only MXNet but also CNTK and TensorFlow, plus other good toolkits that we have not discussed here, including Caffe, Theano, and Torch. Jupyter and the Anaconda tools are there, too.

Configuring the Amazon AMI to use Jupyter is easy. Go to the Amazon Marketplace on the EC2 portal and search for "deep learning"; you will find the Deep Learning AMI. Then select the server type. This AMI is tuned for the p2.16xlarge instances (64 virtual cores plus 16 NVIDIA K80 GPUs). This is an expensive option. If you simply want to experiment, it works well with a no-GPU eight-core option such as m4.2xlarge. When the VM comes up, log in with ssh, and configure Jupyter for remote access as follows.

```
>cd .jupyter
>openssl req -x509 -nodes -days 365 -newkey rsa:1024 \
  -keyout mykey.key -out mycert.pem
>ipython
[1]: from notebook.auth import passwd
[2]: passwd()
Enter password:
Verify password:
Out[2]: 'sha1:---- long string -----------'
```

Remember your password, and copy the long sha1 string. Next create the file .jupyter/jupyter_notebook_config.py, and add the following lines.

```
c = get_config()
c.NotebookApp.password = u'sha1:----long string -----------'
c.NotebookApp.ip = '*'
c.NotebookApp.port = 8888
c.NotebookApp.open_browser = False
```

Now invoke Jupyter as follows.

```
jupyter notebook --certfile=.jupyter/mycert.pem \
                 --keyfile=.jupyter/mykey.key
```

Then, go to https://*ipaddress*:8888 in your browser, where *ipaddress* is the external IP address of your virtual machine. Once you have accessed Jupyter within your browser, visit src/mxnet/example/notebooks to run MXNet.

Many excellent MXNet tutorials are available in addition to those in the AMI notebooks file, for example on the MXNet community site. To illustrate MXNet's use, we examine a particularly deep neural network trained on a dataset with 10 million images. Resnet-152 [152] is a network with 152 convolutional layers based on a concept called deep residual learning, which solved an important problem with training deep networks called the vanishing gradient problem. Put simply, it states that training by gradient descent methods fails for deep networks because as the network grows in depth, the computable gradients become numerically so small that there is no stable descent direction. Deep residual training approaches the problem differently by adding an identity mapping from one layer to the next so that one solves a residual problem rather than the original. It turns out that the residual is much easier for stochastic gradient descent to solve.

Resnets of various sizes have been built with each of the toolkits mentioned here. Here we describe one that is part of the MXNet tutorials [41]. (Notebook 21 provides a Jupyter version.) The network has 150 convolutional layers with a softmax output on a fully connected layer, with 11,221 nodes representing the 11,221 image labels that it is trained to recognize. The input is a $3{\times}224{\times}224$ RGB format image that is loaded into a batch normalization function and then sent to the first convolutional layer. The example first fetches the archived data for the model. There are three main files.

- `resent-152-symbol.json`, a complete description of the network as a large json file
- `resnet-152-0000.params`, a binary file containing all parameters for the trained model
- `synset.txt`, a text file containing the 1,121 image labels, one per line

You can then load the pretrained model data, build a model from the data, and apply the model to a JPEG image. (The image must be converted to $3{\times}244{\times}244$ RGB format: see notebook 21.)

```
import mxnet as mx
# 1) Load the pretrained model data
with open('full-synset.txt','r') as f:
    synsets = [l.rstrip() for l in f]
sym,arg_params,aux_params=
    mx.model.load_checkpoint('full-resnet-152',0)
# 2) Build a model from the data
mod = mx.mod.Module(symbol=sym, context=mx.gpu())
mod.bind(for_training=False, data_shapes=[('data', (1,3,224,224))])
mod.set_params(arg_params, aux_params)
# 3) Send JPEG image to network for prediction
mod.forward(Batch([mx.nd.array(img)]))
prob = mod.get_outputs()[0].asnumpy()
prob = np.squeeze(prob)
a = np.argsort(prob)[::-1]
for i in a[0:5]:
    print('probability=%f, class=%s' %(prob[i], synsets[i]))
```

You will find that the accuracy of the network in recognizing images is excellent. Below we selected four images in figure 10.10 from the Bing image pages with a focus on biology. You can see from the results in table 10.4 that the top choice of the network was correct for each example, although the confidence was less high for yeast and seahorse. These results clearly illustrate the potential for automatic image recognition in aiding scientific tasks.

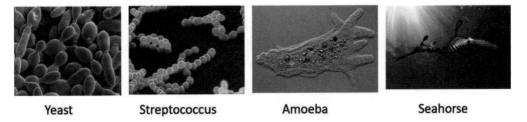

Yeast   **Streptococcus**   **Amoeba**   **Seahorse**

Figure 10.10: Three sample images that we have fed to the MXNet Resnet-152 network.

Table 10.4: Identification of images in figure 10.10 along with estimated probabilities.

Yeast	Streptococcus	Amoeba	Seahorse
p=0.26, yeast	p=0.75, streptococcus, streptococci, strep	p=0.70, ameba, amoeba	p=0.33, seahorse
p=0.21, microorganism	p=0.08, staphylococcus, staph	p=0.15, microorganism	p=0.12, marine animal, marine creature, sea animal
p=0.21, cell	p=0.06, yeast	p=0.05, ciliate, ciliated protozoan, ciliophoran	p=0.12, benthos
p=0.06, streptococcus, strep	p=0.04, microorganism, micro-organism	p=0.04, paramecium, paramecia	p=0.05, invertebrate
p=0.05, eukaryote, eucaryote	p=0.01, cytomegalovirus, CMV	p=0.03, photomicrograph	p=0.04, pipefish, needlefish

10.6   Google TensorFlow in the Cloud

Google's **TensorFlow** is a frequently discussed and used deep learning toolkit. If you have installed the Amazon Deep Learning AMI, then you already have TensorFlow installed, and you can begin experimenting right away. While we have already introduced TensorFlow when discussing convolutional neural networks, we need to look at some core concepts before we dive more deeply.

Let us start with tensors, which are generalizations of arrays to dimensions beyond 1 and 2. In TensorFlow, tensors are created and stored in container objects

Table 10.5: (Fake) graduate school admission data.

GRE	GPA	Rank	Decision
800	4.0	4	0
339	2.0	1	1
750	3.9	1	1
800	4.0	2	0

that are one of three types: variables, placeholders, and constants. To illustrate the use of TensorFlow, we build a logistic regression model of some (fake) graduate school admissions decisions. Our data, shown in table 10.5, consist of a GRE exam score in the (pre-2012) range of 0 to 800; a grade point average (GPA) in the range 0.0 to 4.0; and the rank of the student's undergraduate institution from 4 to 1 (top). The admission decision is binary.

To build the model, we first initialize TensorFlow for an interactive session and define two variables and two placeholders, as follows.

```
import tensorflow as tf
import numpy as np
import csv
sess = tf.InteractiveSession()

x = tf.placeholder(tf.float32, shape=(None,3))
y = tf.placeholder(tf.float32, shape =(None,1))

# Set model weights
W = tf.Variable(tf.zeros([3, 1]))
b = tf.Variable(tf.zeros([1]))
```

The placeholder tensor x represents the triple $[GRE, GPA, Rank]$ from our data and the placeholder y holds the corresponding Admissions Decision. W and b are the learned variables that minimize the cost function.

$$cost = \sum_{i=0}^{1}(y - \sigma(W \cdot x + b))^2)$$

In this equation, $W \cdot x$ is the dot product, but the placeholders are of shape (None, 3) and (None,1), respectively. In TensorFlow, this means that they can hold an array of size N x 3 and N x 1, respectively, for any value of N. The minimization step in TensorFlow now takes the following form, defining a graph with inputs x and y feeding into a cost function, which is then passed to the optimizer to select the W and b that minimize the cost.

```
pred = tf.sigmoid(tf.matmul(x, W) + b)
cost = tf.sqrt(tf.reduce_sum((y - pred)**2/batch_size))
opt = tf.train.AdamOptimizer()
optimizer = opt.minimize(cost)
```

The standard way to train a system in TensorFlow (and indeed in the other packages that we discuss here) is to run the optimizer with successive batches of training data. To do this, we need to initialize the TensorFlow variables with the current interactive session. We use a Python function `get_batch()` that pulls a batch of values from `train_data` and stores them in `train_label` arrays.

```
training_epochs = 100000
batch_size = 100
display_step = 1000
init = tf.initialize_all_variables()

sess.run(init)
# Training cycle
for epoch in range(training_epochs):
    avg_cost = 0.
    total_batch = int(len(train_data)/batch_size)
    # Loop over all batches
    for i in range(total_batch):
        batch_xs,batch_ys=get_batch(batch_size,train_data,train_label)
        # Fit training using batch data
        _,c=sess.run([optimizer,cost],
                        feed_dict={x:batch_xs, y:batch_ys})
        # Compute average loss
        avg_cost += c / total_batch
    # Display logs per epoch step
    if (epoch+1) % display_step == 0:
        print("Epoch:", '%04d' % (epoch+1), "cost=", str(avg_cost))
```

Figure 10.11: TensorFlow code for training the simple logistic regression function.

The code segment in figure 10.11 illustrates how data are passed to the computation graph for evaluation with the `sess.run()` function via a Python dictionary, which binds the data to the specific TensorFlow placeholders. Notebook 23 provides additional details, including analysis of the results. You will see that training on the fake admissions dataset led to a model in which the decision to admit is based solely on the student graduating from the top school. In this case the training rapidly converged, since this rule is easy to "learn." The score was 99.9% accurate. If we base the admission decision on the equally inappropriate policy of granting admission only to those students who either scored an 800 on the GRE or came from the top school, the learning does not converge as fast and the best we could achieve was 83% accuracy.

A good exercise for you to try would be to convert this model to a neural network with one or more hidden layers. See whether you can improve the result!

10.7 Microsoft Cognitive Toolkit

We introduced the **Microsoft Cognitive Toolkit** in section 10.4.3 on page 209, when discussing recurrent neural networks. The CNTK team has made this software available for download in a variety of formats so that deep learning examples can be run on Azure as clusters of Docker containers, in the following configurations:

- CNTK-CPU-InfiniBand-IntelMPI for execution across multiple InfiniBand RDMA VMs

- CNTK-CPU-OpenMPI for multi-instance VMs

- CNTK-GPU-OpenMPI for multiple GPU-equipped servers such as the NC class, which have 24 cores and 4 K80 NVIDIA GPUs

These deployments each use the Azure Batch Shipyard Docker model, part of Azure Batch [6]. (Shipyard also provides scripts to provision Dockerized clusters for MXNet and TensorFlow with similar configurations.)

You also can deploy CNTK on your Windows 10 PC or in a VM running in any cloud. We provide detailed deployment instructions in notebook 22, along with an example that we describe below. The style of computing is similar to Spark, TensorFlow, and others that we have looked at. We use Python to build a flow graph of computations that we invoke with data using an `eval` operation. To illustrate the style, we create three tensors to hold the input values to a graph and then tie those tensors to the matrix-multiply operator and vector addition.

```
import numpy as np
import cntk
X = cntk.input_variable((1,2))
M = cntk.input_variable((2,3))
B = cntk.input_variable((1,3))
Y = cntk.times(X,M)+B
```

X is a 1×2-dimensional tensor, that is, a vector of length 2; M is a 2×3 matrix; and B is a vector of length 3. The expression Y=X*M+B yields a vector of length 3.

However, no computation has taken place at this point: we have only constructed a graph of the computation. To execute the graph, we input values for X, B, and M, and then apply the `eval` operator on Y, as follows. We use Numpy arrays to initialize the tensors and, in a manner identical to TensorFlow, supply a dictionary of bindings to the eval operator as follows.

```
x = [[ np.asarray([[40,50]]) ]]
m = [[ np.asarray([[1, 2, 3], [4, 5, 6]]) ]]
b = [[ np.asarray([1., 1., 1.])]]

print(Y.eval({X:x, M: m, B: b}))

----- output -------------

array([[[[ 241.,   331.,   421.]]]], dtype=float32)
```

CNTK also supports several other tensor container types, such as `Constant`, for a scalar, vector, or other multidimensional tensor with values that do not change, and `ParameterTensor`, for a tensor variable whose value is to be modified during network training.

Many more tensor operators exist, and we cannot discuss them all here. However, one important class is the set of operators that can be used to build multilevel neural networks. Called the *layers library*, they form a critical part of CNTK. One of the most basic is the `Dense(dim)` layer, which creates a fully connected layer of output dimension `dim`. Many other standard layer types exist, including Convolutional, MaxPooling, AveragePooling, and LSTM. Layers can also be stacked with a simple operator called `sequential`. We show two examples taken directly from the CNTK documentation [27]. The first is a standard five-level image recognition network based on convolutional layers.

```
with default_options(activation=relu):
  conv_net = Sequential ([
    # 3 layers of convolution and dimension reduction by pooling
    Convolution((5,5),32,pad=True),MaxPooling((3,3),strides=(2,2)),
    Convolution((5,5),32,pad=True),MaxPooling((3,3),strides=(2,2)),
    Convolution((5,5),64,pad=True),MaxPooling((3,3),strides=(2,2)),
    # 2 dense layers for classification
    Dense(64),
    Dense(10, activation=None)
  ])
```

The second example, on the next page, is a recurrent LSTM network that takes words **embedded** in a vector of size 150, passes them to the LSTM, and produces output through a dense network of dimension `labelDim`.

```
model = Sequential ([
    Embedding(150),         # Embed into a 150-dimensional vector
    Recurrence(LSTM(300)),  # Forward LSTM
    Dense(labelDim)         # Word-wise classification
])
```

You use word embeddings when your inputs are sparse vectors of size equal to the word vocabulary (i.e., if item i in the vector is 1, then the word is the ith element of the vocabulary), in which case the embedding matrix has size vocabulary-size by number of inputs. For example, if there are 10,000 words in the vocabulary and you have 150 inputs, then the matrix is 10,000 rows of length 150, and the ith word in the vocabulary corresponds to the ith row. The embedding matrix may be passed as a parameter or learned as part of training. We illustrate its use with a detailed example later in this chapter.

The Sequential operator used in the same code can be thought of as a concatenation of the layers in the given sequence. The Recurrence operator is used to wrap the correct LSTM output back to the input for the next input to the network. For details, we refer you to the tutorials provided by CNTK. One example of particular interest concerns **reinforced learning**, a technique that allows networks to use feedback from dynamical systems, such as games, in order to learn how to control them. We reference a more detailed discussion online [134].

Azure also provides a large collection of pretrained machine learning services similar to those provided by the Amazon Machine Learning platform: the **Cortana cognitive services**. Specifically, these include web service APIs for speech and language understanding; text analysis; language translation; face recognition and attitude analysis; and search over Microsoft's academic research database and graph. Figure 10.12 shows an example of their use.

10.8   Summary

We have introduced a variety of cloud and open source machine learning tools. We began with a simple logistic regression demonstration that used the machine learning tools in Spark running in an Azure HDInsight cluster. We next turned to the Azure Machine Learning workspace Azure ML, a portal-based tool that provides a drop-and-drag way to compose, train, and test a machine learning model and then convert it automatically into a web service. Amazon also provides a portal-based tool, Amazon Machine Learning, that allows you to build and train a predictive model and deploy it as a service. In addition, both Azure and Amazon provide pre-trained models for image and text analysis, in the Cortana services and the Amazon ML platform, respectively.

We devoted the remainder of this chapter to looking at deep learning and the TensorFlow, CNTK, and MXNet toolkits. The capabilities of these tools can sometimes seem almost miraculous, but as Oren Etzioni [118] observes, "Deep learning isn't a dangerous magic genie. It's just math." We presented a modest introduction to the topic and described two of the most commonly used networks: convolutional and recurrent. We described the use of the Amazon virtual machine image (AMI) for machine learning, which includes MXNet, Amazon's preferred deep learning toolkit, as well as deployments of all the other deep learning frameworks. We illustrated MXNet with the Resnet-152 image recognition network first designed by Microsoft Research. Resnet-152 consists of 152 layers, and we demonstrated how it can be used to help classify biological samples. This type of image recognition has been used successfully in scientific studies ranging from protein structure to galaxy classification [180, 60, 264, 111].

We also used the Amazon ML AMI to demonstrate TensorFlow, Google's open

```
[
  {
    "faceRectangle": {
      "left": 45,
      "top": 48,
      "width": 62,
      "height": 62
    },
    "scores": {
      "anger": 0.0000115756638,
      "contempt": 0.00005204394,
      "disgust": 0.0000272641719,
      "fear": 9.037577e-8,
      "happiness": 0.998033762,
      "neutral": 0.00184232311,
      "sadness": 0.0000301841555,
      "surprise": 0.00000277762956
    }
  }
]
```

Figure 10.12: Cortana face recognition and attitude analysis web service. When applied to an image of a person on a sailboat, it returns the JSON document on the right. Cortana determines that there is one extremely (99.8%!) happy face in the picture.

source contribution to the deep learning world. We illustrated how one defines a convolution neural network in TensorFlow as part of our discussion of that topic, and we provided a complete example of using TensorFlow for logistic regression. Microsoft's cognitive tool kit (CNTK) was the third toolkit that we presented. We illustrated some of its basic features, including its use for deep learning. CNTK also provides an excellent environment for Jupyter, as well as many good tutorials.

We have provided in this chapter only a small introduction to the subject of machine learning. In addition to the deep learning toolkits mentioned here, Theano [47] and Caffe [161] are widely used. Keras `keras.io` is another interesting Python library that runs on top of Theano and TensorFlow. We also have not discussed the work done by IBM with their impressive Watson services—or systems such as Torch `torch.ch`.

Deep learning has had a profound impact on the technical directions of each of the major cloud vendors. The role of deep neural networks in science is still small, but we expect it to grow.

Another topic that we have not addressed in this chapter is the performance of ML toolkits for various tasks. In chapter 7 we discussed the various ways by which a computation can be scaled to solve bigger problems. One approach is the SPMD model of communicating sequential processes by using the Message Passing Standard (MPI) model (see section 7.2 on page 97). Another is the graph execution dataflow model (see chapter 9), used in Spark, Flink, and the deep learning toolkits described here.

Clearly we can write ML algorithms using either MPI or Spark. We should therefore be concerned about understanding the relative performance and programmability of the two approaches. Kamburugamuve et al. [166] address this topic and demonstrate that MPI implementations of two standard ML algorithms perform much better than the versions in Spark and Flink. Often the differences were factors of orders of magnitude in execution time. They also acknowledge that the MPI versions were harder to program than the Spark versions. The same team has released a library of MPI tools called SPIDAL, designed to perform data analytics on HPC clusters [116].

10.9   Resources

The classic *Data Mining: Concepts and Techniques* [148], recently updated, provides a strong introduction to data mining and knowledge discovery. *Deep Learning* [143] is an exceptional treatment of that technology.

For those interested in learning more of the basics of machine learning with Python and Jupyter, two good books are *Python Machine Learning* [224] and *Introduction to Machine Learning with Python: A Guide for Data Scientists* [207]. All the examples in this chapter, with the exception of k-means, involve supervised learning. These books treat the subject of unsupervised learning in more depth.

On the topic of deep learning, each of the three toolkits covered in this chapter— CNTK, TensorFlow, and MXNet—provides extensive tutorials in their standard distributions, when downloaded and installed.

We also mention the six notebooks introduced in this chapter.

- Notebook 18 demonstrates the use of Spark machine learning for logistic regression.

- Notebook 19 can be used to send data to an AzureML web service.

- Notebook 20 demonstrates how to load and use the RNN model originally built with CNTK.

- Notebook 21 shows how to load and use the MXNet Resnet-152 model to classify images.

- Notebook 22 discusses the installation and use of CNTK.

- Notebook 23 illustrates simple logistic regression using TensorFlow.

Chapter 11

The Globus Research Data Management Platform

"Give me where to stand, and I will move the earth."

—Archimedes

We have seen how powerful cloud-based data storage and analysis services can simplify working with large data. But not all science and engineering data live in the cloud. Research is highly collaborative and distributed, and frequently requires specialized resources: data stores, supercomputers, instruments. Thus data are created, consumed, and stored in a variety of locations, including specialized scientific laboratories, national facilities, and institutional computer centers. **Data movement and sharing** and **authentication and authorization** are perennial challenges that can impose considerable friction on research and collaboration.

We describe in this chapter a set of platform services that address these challenges. The Globus cloud service provides data movement, data sharing, and credential and identity management capabilities. We described briefly in section 3.6 on page 51 how these services can be accessed as software as a service, via web interfaces. Here, we introduce more details on these services and describe the Python SDKs that permit their use from within applications. We focus in particular on how the Globus Auth service makes it straightforward to build science services that can accept identities from different identity providers, use standard protocols for authentication and authorization, and thus integrate naturally into a global ecosystem of service providers and consumers. As a use case, we show how these capabilities can be used to build research data portals.

11.1 Challenges and Opportunities of Distributed Data

Data movement is central to many research activities, including analysis, collaboration, publication, and data preservation. However, given its importance and ubiquity, this task remains surprisingly challenging in practice: storage systems have different security configurations, achieving good transfer performance is non-trivial, and as data sizes increase the likelihood of errors increases. Scientists and engineers frequently struggle with such mundane tasks as authenticating and authorizing user access to storage systems, establishing high-speed data connections, and recovering from faults while a transfer proceeds.

Authentication and authorization are similarly central to science and engineering, and for related reasons. Researchers often find themselves needing to navigate a complex world of different identities, authentication methods, and credentials as they access resources in different locations. For example, say that you need to transfer data repeatedly from two sites A and B to a storage system at your home institution H. You have accounts at A and H, with identities U_A and U_H; site B will accept your home institution identity, thanks to the InCommon identity management federation [68]. You will commonly need to authenticate once for each transfer: a painful process and one that prevents scripting. You would prefer to instead authenticate as U_A and U_H just once, and then perform subsequent transfers from A and B to H without further authentications.

The data sharing problem sits at the intersection of these two challenges. Say you want to grant a collaborator access to data at your home institution. Setting up an account just for that purpose is typically a time consuming process, if it is possible at all. And it forces your collaborator to deal with yet another username and password. You need to be able to enable access without a local account.

Globus services address these and other related challenges that arise when our work requires the integration of resources across different locations. As well as an easy-to-use, web-browser based interface, Globus provides REST APIs and Python SDKs to enable the integration of Globus solutions into applications in ways that reduce development costs and increase security, performance, and reliability.

11.2 The Globus Platform

Globus was first introduced in 2010 as a software-as-a-service solution to the problem of moving data between pairs of storage systems or **endpoints** [62, 123]. (An endpoint is a storage system that has been connected to the Globus cloud services by using software called **Globus Connect**.) The Amazon-hosted Globus software handles the complexity involved in transfers, such as authenticating and authorizing user access to endpoints, creating a high-speed data connection between endpoints, and recovering from faults while a transfer proceeds. Importantly, it implements a third-party transfer model in which no data are transferred via the Globus service: instead, data are transferred directly between endpoint pairs by using a protocol called GridFTP that provides specialized support for high performance and reliability [61]. Globus can also perform rsync-like updates when doing repeated transfers, allowing transfer of only new or modified files from the source to the destination. Direct HTTPS transfers to and from endpoints are also supported, allowing web browser access to data stored on Globus endpoints.

The Globus team has subsequently built on this initial **Globus Transfer** service by adding **Auth** for identity and credential management, **Groups** for group management, **Sharing** for data sharing, and **Publication** and **Data Search** for data management. Importantly, the Globus team also created REST APIs and Python SDKs to allow these capabilities to be used programmatically, from within applications. It is these platform capabilities that we describe in this chapter, building on the introductory material in section 3.6.1 on page 52, where we showed how to use the Globus Python SDK to initiate, monitor, and control data transfers. We first provide additional details on the programmatic use of Globus Sharing capabilities, then introduce the use of Globus Auth, and finally present illustrative examples of the use of these capabilities.

11.2.1   Globus Transfer and Sharing

We introduced Globus Sharing capabilities in section 3.6.2 on page 54. Here we show how to use the Python SDK to manage sharing programmatically. Recall that Globus Sharing allows a user to make a specified folder on a Globus endpoint accessible to other Globus users. Figure 11.1 shows the idea. *Bob* has enabled sharing of folder `~/shared_dir` on *Regular* endpoint by creating *Shared endpoint*, and then granting *Jane* access to that shared endpoint. Jane can then use Globus Transfer to read and/or write files in the shared folder, depending on what rights she has been granted.

As is the case with the Globus Transfer service presented in chapter 3, all data sharing capabilities offered by the Globus web interface are also accessible via the Python SDK. The code in figure 11.2 on page 229 illustrates their use. We explain each of the two functions in the figure in turn. We use both functions in section 11.5.3 on page 247 as part of a research data portal implementation.

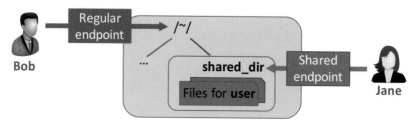

Figure 11.1: The Globus shared endpoint construct allows an authorized administrator of a Globus endpoint (say Bob) to create a **shared endpoint** granting access to a folder within that endpoint, to which they can then authorize access by others (say Jane).

First, the `create_share` function: We assume that we have previously initiated a transfer object, `tc`, in the manner illustrated in the first lines of figure 3.9 on page 55, and that this object is passed to the function, along with the endpoint identifier and path for the folder that is to be shared ("Regular endpoint" and `~/shared_dir`, respectively, in figure 11.1). The function uses the Globus SDK function `operation_mkdir` to request creation of the specified `folder` on the specified `endpoint`. It then creates a parameter structure, calls the Globus SDK function `create_shared_endpoint` to create a shared endpoint for the new directory, and finally returns the identifier for the new endpoint.

Second, the `grant_access` function: This function requires both `tc` and a Globus Auth client reference, `ac` (we introduce Auth in the next section); identifiers for the shared endpoint (`share_id`) for which sharing is to be enabled and the user (`user_id`: a UUID, as with endpoint identifiers) to whom access is to be granted; the type of access to be granted (`atype`: can be `'r'`, `'w'`, or `'rw'`); and a message to be emailed to the user upon completion. The function uses the Globus Auth SDK function `get_identities` to determine the identities that are associated with the user for whom sharing is to be enabled, and extracts from this list an email address. It then uses the Globus Transfer SDK function `add_endpoint_acl_rule` to add an access control rule to the shared endpoint, granting the specified access type to the specified user.

11.2.2 The `rule_data` Structure

Our example program passes a `rule_data` structure to the `add_endpoint_acl_rule` function. The various elements specify, among other things:

- `'principal_type'`: the type of principal to which the rule applies;

```
#
# Create a shared endpoint on specified 'endpoint' and 'folder';
# Return the endpoint id for new endpoint.
# Supplied 'tc' is Globus transfer client reference.
#
def create_share(tc, endpoint, folder):
    # Create directory to be shared
    tc.operation_mkdir(endpoint, path=folder)

    # Create the shared endpoint on specified folder
    shared_ep_data = {
        'DATA_TYPE'     : 'shared_endpoint',
        'host_endpoint': endpoint,
        'host_path'    : folder,
        'display_name' : 'Share ' + folder,
        'description'  : 'New shared endpoint'
    }
    r = tc.create_shared_endpoint(shared_ep_data)
    # Return identifier of the newly created shared endpoint
    return(r['id'])


#
# Grant 'user_id' access 'atype' on 'share_id'; email 'message'
# Supplied 'tc' and 'ac' are Globus Transfer and Auth client refs.
#
def grant_access(tc, ac, share_id, user_id, atype, message):
    # (1)
    r = ac.get_identities(ids=user_id)
    email = r['identities'][0]['email']
    rule_data = {
        'DATA_TYPE'      : 'access',
        'principal_type': 'identity', # To whom is access granted?
        'principal'     : user_id,    # To an individual user
        'path'          : '/',        # Grant access to this path
        'permissions'   : atype,      # Grant specified access
        'notify_email'  : email,      # Email invite to this address
        'notify_message': message     # Include this message in email
    }
    r = tc.add_endpoint_acl_rule(share_id, rule_data)
    return(r)
```

Figure 11.2: A function that uses the Globus Python SDK to create a shared endpoint.

- `'principal'`: as the `'principal_type'` is `'identity'`, this is the user id with whom sharing is to be enabled;

- `'permissions'`: the type of access being granted: in this case read-only (`'r'`), but could also be read and write (`'rw'`);

- `'notify_email'`: an email address to which an invitation to access the shared endpoint should be sent; and

- `'notify_message'`: a message to include in the invitation email.

The `'principal_type'` element can also take the value `'group'`, in which case the `'principal'` element must be a group id. Alternatively, it can take the values `'all_authenticated_users'` or `'anonymous'`, in which cases the `'principal'` element must be an empty string.

11.3 Identity and Credential Management

We noted above the challenges that users face when authenticating to different sites and services in the course of their work. Similarly, service developers need mechanisms for establishing the identity of a requesting user and for determining what that user is authorized to do. Figure 11.3 illustrates some of the concepts and issues involved. An end user wants to run an application that makes requests to remote services on her behalf. Those remote services may themselves want to make further calls to other **dependent services**. For consistency with commonly used terminology, we refer to the user as the **resource owner**, the application as the **client**, and each remote and dependent service as a **resource server**.

Two interrelated problems frequently arise in such contexts. The first concerns the use of **alternative identity providers**. A resource server frequently wants to establish the identity of the user (i.e., resource owner) who issued an incoming request, often to determine whether to grant access and sometimes simply to log who is using their service. In the past, developers of resource servers often implemented their own username-password authentication systems, but such approaches are inconvenient and insecure. Instead, we want to allow a resource server to accept credentials from other identity providers: for example, that associated with a user's home institution. Furthermore, different resource servers may require different credentials. For example, to transfer a file from the University of Chicago to Lawrence Berkeley National Laboratory, I must authenticate with both my Chicago and my Berkeley identities to establish my credentials to access file systems at Chicago and Berkeley, respectively.

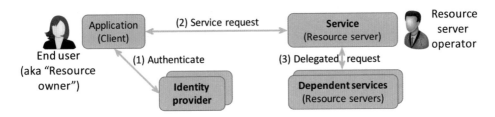

Figure 11.3: A schematic of the entities and interactions that engage in distributed resource accesses, using the terminology of OAuth2.

The second problem concerns **(restricted) delegation**. A resource server may need to perform actions on behalf of a requesting user. For example, it may need to transfer files or perform computations. It may then need credentials that allow it to establish its authority to perform such actions. (This requirement is especially important if the resource server needs to operate in an unattended manner, for example so that it can continue file transfers or computations while the user eats lunch.) However, users may not want to grant unlimited rights to a remote service to perform actions on their behalf, due to the potential for harm if a credential is compromised. Thus, the ability to restrict the rights that are delegated is important. For example, you might be ok with a service reading, but not writing, files on a certain server. And you certainly do not want a compromised service to be able to use other services that you have not authorized.

As we describe in the following, the cloud-hosted Globus Auth service addresses these and other related concerns.

11.3.1 Globus Auth Is an Authorization Service

Globus Auth leverages two widely used web standards, the OAuth 2.0 Authorization Framework (OAuth2) [149] and OpenID Connect Core 1.0 (OIDC) [230], to implement solutions to these problems. OAuth2 is a widely used protocol that applications can use to provide client applications with **secure delegated access**: the delegation that we spoke about above. It works over HTTP and uses **access tokens** to authorize servers, applications, and other entities. OIDC is a simple identity layer on top of the OAuth protocol.

The cloud-hosted Globus Auth service is what OAuth2 calls an **authorization server**. As such, it can issue access tokens to a **client** after successfully authenticating the **resource owner** and obtaining authorization from that resource owner for the client to access resources provided by a **resource server**. (This

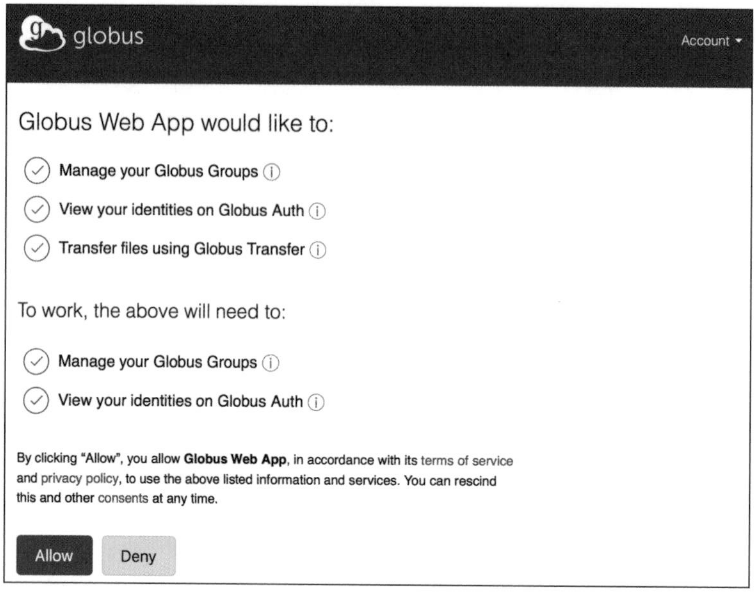

Figure 11.4: A Globus Auth consent request, in this case for the Globus web application.

authorization process typically involves a request for consent, such as those shown in figure 11.4.) The resource owner in this scenario is typically an end user, who authenticates to a Globus Auth-managed Globus account using an identity issued by one of an extensible set of (federated) identity providers supported by Globus Auth. A resource owner could also be a robot, agent, or service acting on its own behalf, rather than on behalf of a user; the client may be either an application (e.g., web, mobile, desktop, command line) or another service acting as a client, as we explain in subsequent discussion.

Having obtained an access token, the client can then present that token as part of a request to the resource server for which the token applies, to demonstrate that it is authorized to make the request. The token is included in the request via the HTTPS Authorization header.

Access tokens are thus the key to OAuth2 and Globus Auth. An access token represents an authorization issued by a resource owner to a client, authorizing the client to request access to a specified resource server on the resource owner's behalf. As we describe later, the resource server can then ask the Globus Auth authorization service for details on what rights have been granted: a process that is referred to as "introspection." For example, if the resource owner in figure 11.3 wants to allow a client (e.g., a web portal) to access a remote service but only for purposes of reading during the next hour, introspection of the associated token can reveal those restrictions. Globus Auth thus addresses the problems of *(restricted) delegation.* It also supports the linking of multiple identities, as we discuss below, to address the problem of *alternative identity providers.*

A resource server receiving a token from a client can thus determine that the resource owner has authorized it to perform certain actions on the resource owner's behalf. What if the resource server then wants to reach out to other resource servers, for example to Globus Transfer to request a data transfer? A problem arises: the resource server has a token that authorizes it to perform actions itself, but it has no token that it can present to the Globus Transfer service to demonstrate that the resource owner (the end user in our example) has authorized transfers.

This is where **dependent services** come in. When a resource server R is registered with Globus Auth, it can specify services that it needs to access to perform its functions: its dependent services, say S and T. A request from R to Globus Auth for authorization then causes Globus Auth to request consent from the user not only for R but also for the dependent services S and T. We saw an example of this scenario in figure 11.4: the Globus web application has registered Globus Transfer and Globus Groups as dependent services, and thus you see the user being asked to consent to those uses. Once consent has been granted, the resource server can request additional dependent access tokens, as required, that it can then include in requests to other services that it makes on the authorizing resource owner's behalf.

OAuth2 and Globus Auth incorporate various complexities and subtleties, but the basic steps are simple. A user accesses an application; Globus Auth authenticates and requests consents from the end user; Globus Auth provides access tokens to the application; the application uses access tokens to access other services; a service receiving an access token can validate it and use it to request dependent access tokens to access other services. Importantly, different actors can play different roles at different times: your web browser can be a client to a web service, that itself can act as a client to other services, and so on.

11.3.2   A Typical Globus Auth Workflow

We use figure 11.5 on page 235 to illustrate how Globus Auth works. The figure looks complicated, but please bear with us: the underlying concepts are straightforward. We describe each of the 12 steps shown in the figure in turn.

1. The end user accesses the application to make a request to a remote service.

The application might be a Web client or, alternatively, an application running on the user's desktop or some other computer.

2. The application contacts Globus Auth to request authorization for the use of a set of **scopes**. A scope represents a set of capabilities provided by a resource server for which an access token is to be granted. In this case, the application requests two scopes: one for access to login information and one for HTTPS/REST API access.

3. Globus Auth arranges for authentication of the user, using an identity provider that is mutually acceptable to the user and the application. Because the user only authenticates with the authorization server, the user's credentials are never shared with the client or with Globus Auth.

4. Globus Auth returns an **authorization code** to the user.

5. The user requests access tokens from Globus Auth, passing the previously acquired authorization code to establish their right to obtain these tokens.

6. Access tokens are returned, one per requested scope. The issuance of multiple tokens enhances security by limiting the impact of a compromise.

7. The client can now use the access token in an HTTPS/REST request to a resource server, by setting an HTTPS `Authorization: Bearer` header with the appropriate token. (For concreteness, the remote service is here shown as Globus Transfer, but it could be anything.)

8. Using a recent OAuth2 extension [226], the resource server can contact Globus Auth to "introspect" the token and thus obtain answers to questions such as "is the token valid?," "which resource owner is it for?," "what client is making the request?," and "which scope is it for?"

9. Globus Auth responds to the introspection request. The resource server can use the provided information to make an authorization decision as to how it responds to the client request.

10. The resource server can also use its access token to request dependent access tokens for any dependent services. For example, Globus Transfer can retrieve an access token for the Globus Groups resource server, so that it can check if the requesting user is a member of a particular group before taking some action like allowing access to a shared endpoint.

11. Globus Auth returns requested dependent tokens.

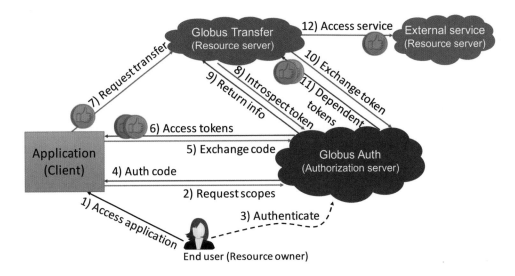

Figure 11.5: Entities and interactions involved in Globus Auth-mediated distributed resource requests. Details are provided in the text.

12. The resource server uses a newly issued dependent access token in an HTTP-S/REST request to the second resource server.

There are other OAuth2 and Globus Auth details that are not covered here: for example, refresh tokens (because an access token's lifetime may be less than that of an application) and the somewhat different methods used in the case of a long-lived application rather than a web browser. Also, an alternative protocol is used for rich clients such as the Javascript-based Globus Transfer client that avoids the needs for steps 4 and 5; a variant of this flow supports mobile, command line, and desktop applications: "native apps." But we have covered the essentials.

11.3.3   Globus Auth Identities

Globus Auth maintains information about the identities that its users may use to authenticate. A Globus Auth identity has a unique, case-insensitive username (for example, `user@example.org`), issued by an identity provider (e.g., a University, research laboratory, or Google), for which a user or client can prove possession via an authentication process (e.g., presenting a password to the identity provider). Globus Auth manages the use of identities (e.g., to login to clients and services), their properties (e.g., contact information), and relationships among identities (e.g., allowing login to an identity by using another linked, "federated" identity).

Globus Auth neither defines its own identity usernames nor verifies authentication (e.g., via passwords) with identities. Rather, it acts as an intermediary between external identity providers, on the one hand, and clients and services that want to leverage identities issued by those providers, on the other. Globus Auth assigns each identity that it encounters an identifier: a UUID that is guaranteed to be unique among all Globus Auth identities, and that will never be reused. This ID is what resource servers and clients should use as the canonical identifier for a Globus Auth identity. Associated with this ID are an identity provider, a username given to the identity by the provider, and other provider-supplied information such as display name and contact email address.

An example Globus Auth identity. The following is an example of the information that may be associated with a Globus Auth identity:

username	: **rocky@wossamotta.edu**
id	: **de305d54-75b4-431b-adb2-eb6b9e546014**
identity_provider	: **wossamotta.edu**
display_name	: **Rocket J. Squirrel**
email	: **rocky@wossamotta.edu**

Globus supports more than 100 identity providers, and more are being added all the time. Examples include the many US and international universities and other institutions that support InCommon; various identity providers that support the OpenID Connect protocol; Google; and the Open Researcher and Contributor ID (ORCID). The process of integrating a new identity provider is beyond the scope of this book, but it is a straightforward process. See the Globus documentation for more information.

11.3.4 Globus Accounts

An identity can be used with Globus Auth to create a **Globus account**. A Globus account has a primary identity, but can also have any number of other identities linked to it as well. Thus, for example, Mr. Squirrel may create a Globus account with the identity above and then link to that account a Google identity, his ORCID, and an identity provided by a scientific facility to which he has access.

A Globus account is not an identity itself. It does not have its own name or identifier. Rather, a Globus account is identified by its primary identity. Similarly, profile information and other metadata are tied to identities, not to accounts. A Globus account is simply a set of identities comprising the primary identity and all identities linked to that primary identity.

11.3.5 Using Globus Auth Identities

Clients and resource servers should always use the Globus Auth-provided identity ID when referring to an identity, for example in access control lists, and when referring to identities in a REST API. Clients and resource servers can use the Globus Auth REST API to map any identity username to its (current) identity ID, and request information about an identity ID (e.g., username, display_name, provider, email), for example as follows:

```
import globus_sdk
# Obtain reference to Globus Auth client
ac = globus_sdk.AuthClient()
# Get identifies associated with username 'globus@globus.org'
id = ac.get_identities(usernames='globus@globus.org')
# Return zero or more UUIDs
# Get identities associated with a UUID
r = ac.get_identities(ids=id)
```

The last command returns a JSON document containing a list of identities, such as the following. (This example document contains just one identity.)

```
{'identities':
    [ { 'email'              : None,
        'id'                 : '46bd0f56-e24f-11e5-a510-131bef46955c',
        'identity_provider': '7daddf46-70c5-45ee-9f0f-7244fe7c8707',
        'name'               : None,
        'organization'       : None,
        'status'             : 'unused',
        'username'           : 'globus@globus.org'}
    ]
}
```

11.3.6 Use of Globus Auth by Resource Servers

Having introduced various details of the Globus Auth server, Globus Auth identities, and Globus accounts, we can now turn to the practical question of what we can do with these mechanisms. In particular, we describe how resource servers can use Globus Auth as an authorization server and thus both support sophisticated OAuth2 and OpenID Connect functionality, and leverage other resource servers that use Globus Auth.

Let us consider, for example, a research data service that accepts user requests to analyze genomic sequence data. (We describe an example of such a system, Globus Genomics, in section 14.4 on page 303.) This service is basically a data and code repository with a REST API, which other applications can leverage to access this repository programmatically.

This service is a resource server in the Globus Auth context. It needs to be able to authenticate users, validate user requests, and make requests to other services (e.g., to cloud or institutional storage to retrieve sequence data and store results, and to computing facilities to perform computations) on a user's behalf. Globus Auth allows us to program each of these capabilities via manipulation of identities, access tokens, and OAuth2 protocol messages.

Assume that some client to this service has already followed steps 1–7 in figure 11.5 on page 235 and thus possesses the necessary access tokens. (The "client" may be a web client to the data server, or some other web, mobile, desktop, or command line application.) Interactions may then proceed as follows.

1. The client makes an HTTPS request to the resource server (the research data service proper: in the following we refer to it as the "data service") with an `Authorization: Bearer` header containing an access token. (Step 8 in figure 11.5.)

2. The data service calls the function `oauth2_token_introspect` provided by the Globus Auth SDK, authorized by the data service's client identifier and client secret (see below), to validate the request access token, and obtain additional information related to that token (scopes, effective identity, identities set, etc.). If the token is not valid, or is not intended for use with this resource server, Globus Auth returns an error.

3. The data service verifies that the request from its client conforms to the scopes associated with the request access token.

4. The data service verifies the identity of the resource owner (typically an end user) on whose behalf the client is acting. The data service may use this identity as its local account identifier for this user.

5. The data service uses the set of identities associated with the account referred to by the request access token to determine what the request is allowed to do. For example, if the request is to access a resource that is shared with particular identities, the data service should compare all of the account's identities (primary and linked identity ids) with the resource access control permissions to determine if the request should be granted.

6. The data service may need to act as a client to other (dependent) resource servers, as discussed above. In that case, the data service uses the Globus SDK

`oauth2_get_dependent_tokens` function to get dependent access tokens for use with downstream resource servers, based on the request access token that it received from the client.

7. The data service uses a dependent access token to make a request to a dependent resource server.

8. The data service responds to its client with an appropriate response.

A note regarding the client identifier and client secret mentioned in Step 2: Each client and resource server must register with Globus Auth and obtain a **client id** and **client secret**, which they can subsequently use with Globus Auth to prove who it is in the various OAuth2 messages: for example, when swapping an authorization token for an access token, calling token introspect, calling dependent token grant, and using a refresh token to obtain a new access token.

11.3.7 Other Globus Capabilities

Globus also supports a growing set of other capabilities beyond those described here. For example, table 11.1 lists additional functions supported by the Globus Transfer Python SDK.

Table 11.1: Some of the close to 50 functions supported by the Globus Transfer Python SDK. (Others mostly implement endpoint administration functions.)

Type	Function	Description
Endpoint information	`endpoint_search`	Search on name, keywords, etc.
	`get_endpoint`	Get endpoint information
	`my_shared_endpoint_list`	Get endpoints that I manage
File system operations	`operation_mkdir`	Create a folder on endpoint
	`operation_ls`	List contents of endpoint
	`operation_rename`	Rename folder or directory
Task management	`submit_transfer`	Submit a transfer request
	`submit_delete`	Submit a delete request
	`cancel_task`	Cancel submitted request
	`task_wait`	Wait for task to complete
Task information	`task_list`	Get information about tasks
	`get_task`	Get information about a task
	`task_event_list`	Get event info for a task
	`task_successful_transfers`	Get successful transfers for task
	`task_pause_info`	Get info on why task paused

Other Globus services provide other capabilities. Globus Publication, for example, provides user-configurable, cloud-hosted data publication pipelines that can be used to automate the workflows used to make data accessible to others, workflows that will typically include steps such as providing and collecting metadata, moving data to long-term storage, assigning persistent identifiers (e.g., a Digital Object Identifier or DOI [218]), and verifying data correctness [89]. Globus Data Search can be used to search for data on endpoints to which a user has access. See the Globus documentation `docs.globus.org` for information on these services.

Data delivery at the Advanced Photon Source. The **Advanced Photon Source** (APS) at Argonne National Laboratory is typical of many experimental facilities worldwide in that it serves large numbers (thousands) of researchers every year, most of whom visit just for a few days to collect data and then return to their home institution. In the past, data produced during an experiment was invariably carried back on physical media. However, as data sizes have grown and experiments have become more collaborative, that approach has become less effective. Data transfer via network is preferred; the challenge is to integrate data transfer into the experimental workflow of the facility in a way that is fully automated, secure, reliable, and scalable to thousands of users and datasets.

Francesco De Carlo uses Globus APIs to do just that at the APS. His **DMagic system** [107] implements a variant of the program in figure 11.9 that integrates with APS administrative and facility systems to deliver data to experimental users. When an experiment is approved at the APS, a set of associated researchers are registered in the APS administrative database as approved participants. DMagic leverages this information as follows. Before the experiment begins, it creates a shared endpoint on a large storage system maintained by Argonne's computing facility. DMagic then retrieves from the APS scheduling system the list of approved users for the experiment, and adds permissions for those users to the shared endpoint. It then monitors the experiment data directory at the APS facility and copies new files automatically to that shared endpoint, from which it can be retrieved by any approved user.

11.4 Building a Remotely Accessible Service

Say you want to build a service that can be invoked remotely via a REST API call. Building and invoking a service in this way is straightforward in principle: many libraries exist for defining, implementing, and using REST APIs. Security is perhaps the one major source of complexity, and here Globus Auth can help. The basic issue is that when a remote user makes a request to the service, the service author needs to be able to determine who is making the request and what rights the requestor is passing with the request. For example, the service may want to know if it is permitted to make Globus transfer requests on behalf of the requesting user. It may also want to know the identity of the requestor, so that the requestor can be given access to a shared endpoint created by the service.

To illustrate how Globus Auth can be used to address these concerns, we present a simple **Graph service** that accepts requests to generate graphs of temperature data. In response to a request, it retrieves data from a web server, generates graphs, and uses Globus Transfer to transfer the graphs to the requestor. It thus needs to authenticate and authorize the requestor and obtain dependent access tokens for a web server and Globus Transfer. A complete Python implementation of this example service is available at `github.com/globus/globus-sample-data-portal`, in the folder `service`. We use extracts (some simplified) from this implementation to illustrate how the Graph service works with Globus Auth.

The relevant authorization code is in figure 11.6 on the next page. The Graph service receives a HTTPS request with a header containing the access token in the form `Authorization: Bearer <request-access-token>`. It then uses the following code to (1) retrieve the access token, (2) call out to Globus Auth to retrieve information about the token, including its validity, client, scope, and effective identity. The Graph service can then (3–5) verify the token information and (6) authorize the request. (In our example, every request is accepted.)

This sample code has been written so that it only (5) accepts requests from an entity that can supply a `PORTAL_CLIENT_ID`, a service that we introduce later in the chapter. As we show in the next paragraph, it then requests and obtains dependent access tokens that allow it to transfer data on behalf of that entity. An alternative implementation, to be preferred if we want the Graph service to be more broadly useful, would have it look for the original resource owner's (end user's) token and then perform operations on their behalf.

As the Graph service needs to act as a client to the data service on which the datasets are located, it next requests dependent tokens from Globus Auth. This and subsequent code fragments in this section are from the file `service/view.py`.

```
# (1) Get the access token from the request
token = get_token(request.headers['Authorization'])

# (2) Introspect token to extract
client = load_auth_client()
token_meta = client.oauth2_token_introspect(token)

# (3) Verify that the token is active
if not token_meta.get('active'):
    raise ForbiddenError()

# (4) Verify that "audience" for this token is our service
if 'Graph Service' not in token_meta.get('aud', []):

raise ForbiddenError()

# (5) Verify that identities_set in token includes portal client
if app.config['PORTAL_CLIENT_ID'] != token_meta.get('sub'):
    raise ForbiddenError()

# (6) Token has passed verification: stash in request global object
g.req_token = token
```

Figure 11.6: Selected and somewhat simplified code from the file `service/decorators.py` in the Graph service example.

```
client = load_auth_client()
dependent_tokens = client.oauth2_get_dependent_tokens(token)
```

Having retrieved these dependent tokens, it extracts from them the two access tokens that allow it to itself act as a client to the Globus Transfer service and to an HTTPS endpoint service from which it will retrieve datasets.

```
transfer_token = dependent_tokens.by_resource_server[
    'transfer.api.globus.org']['access_token']
http_token = dependent_tokens.by_resource_server[
    'tutorial-https-endpoint.globus.org']['access_token']
```

The service also extracts from the request details of the datasets to be graphed, and the identity of the requesting user for use when configuring the shared endpoint:

```
selected_ids = request.form.getlist('datasets')
selected_year = request.form.get('year')
user_identity_id = request.form.get('user_identity_id')
```

The Graph service next fetches each `dataset` via an HTTPS request to the data server, using code like the following. The previously obtained `http_token` provides the credentials required to authenticate to the data server.

```
response = requests.get(dataset,
    headers=dict(Authorization='Bearer ' + http_token))
```

A graph is generated for each dataset. Then, the Globus SDK functions `operation_mkdir` and `add_endpoint_acl_rule` are used, as in section 11.2.1 on page 227, to request that Globus Transfer create a new shared endpoint accessible by the user identity that was previously extracted from the request header, `user_identity_id`. (The `transfer_token` previously obtained from Globus Auth provides the credentials required to authenticate to Globus Transfer.) Finally, the graph files are transferred to the newly created directory via HTTP, using the same `http_token` as previously, and the Graph server sends a response to the requester, specifying the number and location of the graph files.

This example shows how Globus Auth allows you to outsource all identity management and authentication functions. Identities can be provided by federated identity providers, such as InCommon and Google. All REST API security functions, including consent and token issuance, validation, and revocation, are provided by Globus Auth. Your service needs only to provide service-specific authorization, which can be performed on the basis of identity or group membership. And because all interactions are compliant with OAuth2 and OIDC standards, any application that speaks these protocols can use your service as they would any other; your service can seamlessly leverage other services; and other services can leverage your service. You can easily build a service to be made available to others as part of the national cyberinfrastructure; equally, you can build a service that dispatches requests to other elements of that cyberinfrastructure.

11.5   The Research Data Portal Design Pattern

To further illustrate the use of Globus platform services in scientific applications and workflows, we describe how they may be used to realize a design pattern that Eli Dart calls the **research data portal**. In this design pattern, specialized properties of modern research networks are exploited to enable high-speed, secure delivery of data to remote users. In particular, the control logic used to manage data access and delivery is separated from the machinery used to deliver data over high-speed networks. In this way, order-of-magnitude performance improvements can be achieved relative to traditional portal architectures in which control logic and data servers are co-located behind performance-limiting firewalls and on low-performance web servers.

11.5.1 The Vital Role of Science DMZs and DTNs

A growing number of research universities and laboratories worldwide are connected in a network fabric that links data stores, scientific instruments, and computational facilities at unprecedented speeds: 10 or even 100 gigabits per second (Gb/s). Increasingly, research networks are themselves connected to cloud providers at comparable speeds. Thus, in principle, it should be possible to move data between any element of science and engineering infrastructure with great rapidity.

In practice, real transfers often achieve nothing like these theoretically achievable peak speeds. One common reason for poor performance is firewalls or other bottlenecks in the network connection between the outside world and the device from/to which data are to be transferred: the so-called "last mile"—or, outside the US, the last kilometer. The firewalls are often there for a good reason, such as protecting the sensitive data contained on the administrative computers that are also connected to the global Internet. But they get in the way of high-bandwidth science and engineering traffic. The other common reason for poor performance is using tools not designed for performance, like secure copy (SCP).

Two concepts, the Science DMZ and Data Transfer Node, are now being widely deployed to overcome this problem. The **Science DMZ** overcomes the challenges associated with multipurpose enterprise network architectures by placing resources that need high-performance connectivity in a special subnetwork that is close (from a network architecture perspective) to the border router that connects the institution to the high-speed wide area network. Traffic between those resources and the outside world can then bypass internal firewalls.

Note that the point here is not to circumvent security by putting it outside the firewall. Rather, it is about recognizing that there is certain traffic for which firewalls not only slow things down, but are not needed. The Science DMZ uses alternative network security approaches that are appropriate for such traffic. For example, the DTN is not wide open: the Science DMZ router blocks most ports. But the ports necessary for secure, high-performance data transfer are open, and avoid the packet-inspecting firewalls.

A **Data Transfer Node** (DTN) is a specialized device dedicated to data transfer functions. These devices are typically Linux servers constructed with high quality components, configured for both high-speed wide area data transfer and high-speed access to local storage resources. DTNs run the high-performance

Globus Connect data transfer software, introduced in section 3.6 on page 51, to connect their storage to the Globus cloud and thus to the rest of the world. General-purpose computing and business productivity applications, such as email clients and document editors, are not installed; this restriction produces more consistent data transfer behavior and makes security policies easier to enforce.

The Science DMZ design pattern also includes other elements, such as integrated perfSONAR monitoring devices [244] for performance debugging, specialized security configurations, and variants used to integrate supercomputers and other resources. But this brief description covers the essentials. The U.S. Department of Energy's Energy Sciences Network (ESnet) has produced detailed configuration and tuning guides for Science DMZs and DTNs `fasterdata.es.net`.

11.5.2   The Research Data Portal Application

Eli Dart coined the term research data portal to indicate a web service designed primarily to serve data to remote users. (The variant that accepts data from remote users, for example for analysis or publication, has similar properties.) A research data portal must be able to authenticate and authorize remote users, allow those users to browse and query a potentially large collection of data, and return selected data (perhaps after subsetting) to remote users. In other words, a research data portal is like a web server, except that the data that it serves may be orders of magnitude larger than typical web pages.

Figure 11.7 shows how research data portals have often been architected in the past. A single **data portal server** both runs portal logic and serves data from local storage. This architecture is simple but cannot easily achieve high performance. The problem is that the control logic, being concerned with sensitive topics such as authentication and authorization, needs to sit behind the enterprise firewall. But this arrangement means that all data served by the portal also pass through the firewall, which typically means that they are delivered at a small fraction of the theoretical peak performance of the available networks.

As figure 11.8 on the next page shows, Science DMZs and DTNs allows for new architectural approaches that combine high-speed access and secure operations. The basic idea is to separate what we may call the portal **control channel** communications (i.e., those concerned with such tasks as user authentication and data search) and **data channel** communications (i.e., those concerned with data upload and download delivery). The former can be located on a modestly sized web server computer protected by the institution's firewall, with modest capacity networks, while the latter can be performed via high-speed DTNs and can use specialized protocols such as GridFTP. The research data portal design pattern thus defines distinct roles for the web server, which manages who is allowed to do what, and the Science DMZ, where authorized operations are performed.

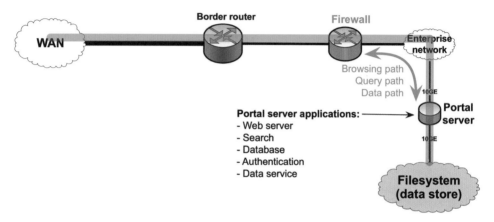

Figure 11.7: A legacy data portal, in which both control traffic (queries, etc.) and data traffic must pass through the enterprise firewall. Figure courtesy Eli Dart.

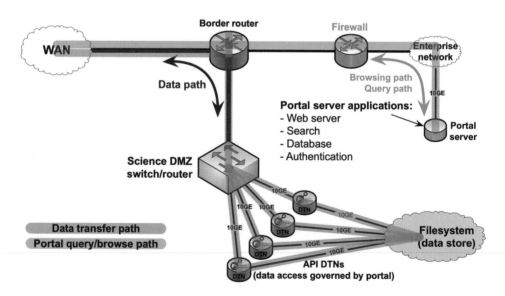

Figure 11.8: A modern research data portal, showing the high-speed data path through the border router and to the DTN in green and the control path through the enterprise firewall to the portal server in red. Multiple DTNs provide for high-speed transfer between network and storage. Figure courtesy Eli Dart.

11.5.3   Implementing the Design Pattern with Globus

We now need mechanisms to allow research code running on the portal server to manage access to, and drive transfers to and from, the DTNs. This is where Globus SDKs come in, as we discuss next. We consider a use case similar to the NCAR Research Data Archive example that follows. A user requests data for download; the portal makes the data available via four steps: (1) create a shared endpoint; (2) copy the requested data to that shared endpoint; (3) set permissions on the shared endpoint to enable access by the requesting user, and email the user a URL that they can use to retrieve data from the shared endpoint; and ultimately (perhaps after several days or weeks), (4) delete the new shared endpoint.

The NCAR Research Data Archive (RDA) [48] `rda.ucar.edu` operated by the U.S. National Center for Atmospheric Research illustrates some of the issues that can arise when implementing a research data portal. This system contains more than 600 data collections, ranging in size from gigabytes to tens of terabytes, and including meteorological and oceanographic observations, operational and reanalysis model outputs, and remote sensing datasets, along with ancillary datasets, such as topography/bathymetry, vegetation, and land use.

The RDA data portal allows users to browse and search catalogs of environmental datasets, place datasets that they wish to download into a "shopping basket," and then download selected datasets to their personal computer or other location. (RDA users are primarily researchers at federal and academic research laboratories. In 2014 alone, more than 11,000 people downloaded more than 1.1 petabytes.) The portal must thus implement a range of different functions, some totally domain-independent (e.g., user identities, authentication, and data transfer) and others more domain-specific (e.g., a catalog of environmental data collections). As we see later in the chapter, the beauty of the Globus approach is that much of the domain-independent logic—in particular, that associated with identity management, authentication, data movement, and data sharing—can be outsourced to cloud services.

We present in figure 11.9 a function `rdp` that implements these actions. As shown in the following, this function takes as arguments the identifier for the endpoint on which the shared endpoint is to be created; the folder on that endpoint for which sharing is to be enabled (here, `Share123`, or `shared_dir` in figure 11.1 on page 228); the folder on that endpoint from which the contents of the shared folder are to be copied; the identifier for the user to be granted access to the new endpoint; and an email address to send a notification of the new share.

```
rdp('b0254878-6d04-11e5-ba46-22000b92c6ec',
    'Share123',
    '~/TEST/',
    'cce13ca1-493a-46e1-a1f0-08bc219638de',
    'foster@anl.gov')
```

As noted in section 3.6 on page 51 and shown in this example, each Globus endpoint and user is named by a universally unique identifier (UUID). An endpoint's identifier can be determined via the Globus web client or programmatically; a user's identifier can be determined programmatically, as we show in notebook 8.

The code in figure 11.9 proceeds as follows. In steps 1 and 2, we obtain Transfer and Auth client references and use `endpoint_autoactivate`, a Globus SDK function, to ensure that the research data portal admin has a credential that permits access to the endpoint identified by `host_id`. (See section 3.6.1 on page 52 for more discussion of `endpoint_autoactivate`.)

In step 3, we call the function `create_share` of figure 11.2 on page 229, passing as parameters the Transfer client reference, the identifier for the endpoint on which the shared endpoint is to be created, and the path for the folder that is to be shared: in our example call, the directory `/~/Share123`. As discussed earlier, that function creates a shared endpoint for the new directory. At this point, the new shared endpoint exists and is associated with this directory. However, only the creating user has access to this new shared endpoint at this point.

In step 4, we use a Globus transfer to copy the contents of the folder `source_path` to the new shared endpoint. (The transfer here is from the endpoint on which the new shared endpoint has been created, but it could be from any Globus endpoint that the research data portal admin is authorized to access.) We have already introduced the Globus Transfer SDK functions used here in section 3.6 on page 51.

In step 5, we call the `grant_access` function defined in figure 11.2 to grant our user access to the new shared endpoint. The function call specifies the type of access to be granted (`'r'`: read only) and the message to be included in a notification email: `'Your data are available'`. The invitation letter sent to the user by the Globus SDK function `add_endpoint_acl_rule` is shown in figure 11.10.

The user is now authorized to download data from the new shared endpoint. That endpoint will typically be left operational for some period, after which it can be deleted, as shown in step 6. Note that deleting a shared endpoint does not delete the data that it contains: The research data portal administrator may want to retain the data for other purposes. If the data are not to be retained, we can use the Globus SDK function `submit_delete` to delete the folder.

```
from globus_sdk import TransferClient, TransferData, AuthClient
import sys, random
def rdp(host_id,      # Endpoint on which to create shared endpoint
        source_path,  # Directory to copy shared data from
        shared_dir,   # Directory name for shared endpoint
        user_id):     # User to share with

    # (1) Obtain Transfer and Auth client references
    tc = TransferClient()
    ac = AuthClient()

    # (2) Activate host endpoint
    tc.endpoint_autoactivate(host_id)

    # (3) Create shared endpoint
    share_id = create_share(tc, host_id, '/~/' + shared_dir + '/')

    # (4) Copy data into the shared endpoint
    tc.endpoint_autoactivate(share_id)
    tdata = TransferData(tc, host_id, share_id,
                label='Copy to share', sync_level='checksum')
    tdata.add_item(source_path, '/~/', recursive=True)
    r = tc.submit_transfer(tdata)
    tc.task_wait(r['task_id'], timeout=1000, polling_interval=10)

    # (5) Set access control to enable access by user
    grant_access(tc, ac, share_id, user_id, 'r',
                'Your data are available')

    # (6) Ultimately, delete the shared endpoint
    tc.delete_endpoint(share_id)
```

Figure 11.9: Globus code to implement research data portal design pattern.

249

From: Globus Notification <noreply@globus.org>
To: Portal server user <user@user.org>
Subject: Portal server admin (admin@therdp.org) shared folder "/" on "Share123" with you

---

Globus user Portal server admin (admin@therdp.org) shared the folder "/" on the endpoint "Share123" (endpoint id: 698062fa-88ed-11e6-b029-22000b92c261) with user@user.org, with the message:

Your data are available.

Use this URL to access the share:
https://www.globus.org/app/transfer?&origin_id=698062fa-88ed-11e6-b029-22000b92c261&origin_path=/&add_identity=cce13ca1-493a-46e1-a1f0-08bc219638de

The Globus Team
support@globus.org

Figure 11.10: Invitation email sent by the program in figure 11.9.

A variant of this approach, with certain administrative advantages, is as follows. Rather than having the portal server create a new shared endpoint for each request, a single shared endpoint is created once and the portal is given the access manager role on the shared endpoint so that it can set ACL rules. Then, for each request it creates a folder on the shared endpoint, puts the data in that location, and sets an ACL rule to manage access. Cleanup is then simpler: the portal just removes the ACL rule and deletes the folder.

11.6 The Portal Design Pattern Revisited

The preceding example shows the essentials of a Globus implementation of the research data portal design pattern. We provide in figure 11.11 a more abstract picture of the architecture that makes clear the components involved and their relationships. To recap, the **portal web server** at the center of the figure is where all custom logic associated with the research data portal sits. This portal server acts as a client, in the Globus Auth/OAuth2 sense, to the other services that it uses to handle the heavy lifting of authentication and authorization (Globus Auth), data transfer and sharing (Globus Transfer), and other computations (Other services). The user accesses portal capabilities via a web browser, and data transfers occur between Globus Connect servers at various locations.

Many variants of this basic research data portal design pattern can be imagined. A minor variant is to prompt the user for where they want their data placed; the portal then submits a transfer on the user's behalf to copy the data to the specified endpoint and path, hence automating yet another step. Or, the data that users access may come from experimental facilities rather than a data archive, in which case data may be deleted after successful download. Access may be granted to groups of users rather than individuals. A portal may allow its users to upload datasets for analysis and then retrieve analysis results. A data publication portal may accept data submissions from users, and load data that pass quality control procedures into a public archive. We give examples of several such variants in the following, and show that each can naturally be expressed in terms of the same basic design pattern.

Similarly, while we have described the research data portal in the context of an institutional Science DMZ, in which (as shown in figure 11.7) the portal server and data store both sit within the research institution, other distributions are also possible and can have advantages. For example, the portal can be deployed on the public cloud for high availability, while the data sits in the Science DMZ to enable direct access from high-speed research networks and/or to avoid public cloud storage charges. Alternatively, the portal can be in the research institution and data in cloud storage. Or both components can be run on cloud resources.

Regardless of the specifics, a research data portal typically needs to perform mundane but important tasks such as determining the identity of a user who wants to access the service; controlling which users are able to access different data and other services within the portal; uploading data reliably, securely, and efficiently from a variety of locations to storage systems within the Science DMZ; downloading data reliably, securely, and efficiently from storage systems within the Science DMZ to a variety of locations; dispatching requests to other services on behalf of users; and logging all actions performed for purposes of audit, accounting, and reporting. Each task is modestly complex to implement and operate reliably and well. Building on top of existing services can not only greatly reduce development costs, but also increase code quality and interoperability via use of standards.

As figure 11.9 shows, the benefits of this approach lie not only in the separation of concerns between control logic and data movement. In addition, the portal developer and admin both benefit from the ability to hand off the management of file access and transfers to the Globus service. The use of Globus APIs makes it easy to implement a wide range of behaviors via simple programs; Globus handles the heavy lifting of high-quality, reliable, and secure authentication, authorization, and data management.

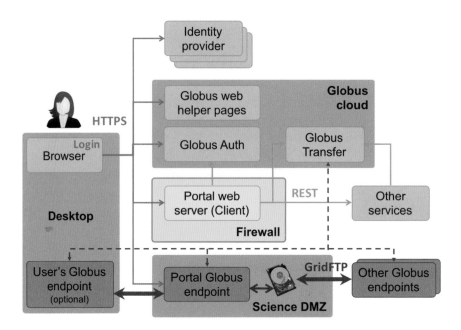

Figure 11.11: The research data portal architecture, showing principal components. Only the portal web server logic needs to be provided by the portal developer. Not shown are other applications that, like the Browser on the left, may access the portal server: for example, command line, thick client, or mobile applications.

It is these capabilities that made it easy to realize the example systems mentioned in this chapter: the NCAR Research Data Archive, which provides high-speed delivery of research data to geoscientists; the DMagic data sharing system for data distribution from light sources; and the Sanger Imputation Service (described on the next page), which supports online analysis of user-provided genomic data.

11.7 Closing the Loop: From Portal to Graph Service

We have already shown in section 11.4 on page 240 how to use the Globus Auth SDK to implement a service that responds to requests from a portal server: the arrow labeled REST from the **Portal web server** to **Other services** in figure 11.11. Such calls might be used in a research data portal for several reasons. You might want to organize your portal as a lightweight front end (e.g., pure Javascript) that interacts with one or more remote backend services. Another reason is that you might want to provide a public REST API for the main portal machinery, so that other app and service developers can integrate with and build on your portal.

Now we look at the logic and code involved in *generating* such requests. Our research data service skeleton illustrates this capability. When a user selects the (Graph) option to request that datasets be graphed, the portal does not perform those graphing operations itself but instead sends a request to a separate Graph service. The request provides the names of the datasets to be graphed. The Graph service retrieves these datasets from a specified location, runs the graphing program, and uploads the resulting graphs to a dynamically created shared endpoint for subsequent retrieval. We describe in the following both the portal server and Graph server code used to implement this behavior.

Figure 11.12 shows a slightly simplified version of the portal code that sets up, sends, and processes the response from the graph request, using the Python Requests library [225]. The code (1) retrieves the access tokens obtained during authentication and extracts the access token for the graph service. (The graph service scope is requested during this flow.) It then (2) assembles the URL, (3) header (containing the Graph service access token), and (4) data for the REST call (including information about the requesting user), and (5) dispatches the call. The remainder of the code (6) checks for a valid response, (7) extracts the location of the newly created graph files from the response, and (8) and directs the user to a Globus transfer browser to access the files.

Sanger Institute Imputation Service `imputation.sanger.ac.uk`. Operated by the Sanger Institute in the UK, this service allows you to upload files containing genome wide association study (GWAS) data from the 23andMe genotyping service and receive back the results of imputation and other analyses that identify genes that you are likely to possess based on those data. The service uses Globus APIs to implement a variant of the research data service design pattern, as follows.

A user who wants to use the service first registers an imputation job. As part of this process, they are prompted for their name, email address, and Globus identity, and the type of analysis to be performed. The Sanger service then requests Globus to create a shared endpoint, share that endpoint with the Globus identity provided by the user, and email a link to this endpoint to the user. The user clicks on that link to upload their GWAS data file and the corresponding imputation task is added to the imputation queue at the Sanger Institute. Once the imputation task is completed, the Sanger service requests Globus to create a second shared endpoint to contain the output and to email the user a link to that new endpoint for download. The overall process differs from that of figure 11.9 only in that a shared endpoint is used for data upload as well as download.

```
# (1) Get access tokens for the Graph service
tokens = get_portal_tokens()
gs_token = tokens.get('Graph Service')['token']

# (2) Assemble URL for REST call
gs_url = '{}/{}'.format(app.config['SERVICE_URL_BASE'], 'api/doit')

# (3) Assemble request headers
req_headers = dict(Authorization= 'Bearer {}'.format(gs_token))

# (4) Assemble request data. Note retrieval of user info.
req_data = dict(datasets=selected_ids,
                year=selected_year,
                user_identity_id=session.get('primary_identity'),
                user_identity_name=session.get('primary_username'))

# (5) Post request to the Graph service
resp = requests.post(gs_url,
                     headers=req_headers,
                     data=req_data,
                     verify=False)

# (6) Check for valid response
resp.raise_for_status()

# (7) Extract information from response
resp_data = resp.json()
dest_ep = resp_data.get('dest_ep')
dest_path = resp_data.get('dest_path')

# (8) Show Globus endpoint browser for new data
return redirect(url_for('browse', endpoint_id=dest_ep,
                        endpoint_path=dest_path.lstrip('/')))
```

Figure 11.12: Slightly simplified version of the function graph() in file portal/view.py at github.com/globus/globus-sample-data-portal.

254

11.8   Summary

In the distributed, collaborative, data-rich world of modern science, the abilities to transfer, share, and analyze data regardless of location, and to navigate complex security regimes in so doing, are frequently essential to progress. We have described cloud-hosted platform services, Globus Auth, Transfer, and Sharing, to which developers of applications and tools that must operate in this world can outsource responsibility for such tasks. We have used the example of a research data portal to illustrate their use, but they can be used in many different configurations.

11.9   Resources

Globus has extensive online documentation for their REST APIs, Python SDK, and command line interface `dev.globus.org`. Dart et al. [106] provide additional information on the Science DMZ concept and design pattern.

Part IV

Building Your Own Cloud

Building your own cloud
What you need to know
Using Eucalyptus
Part IV Using OpenStack

Security and other topics
Securing services and data
Solutions
History, critiques, futures   **Part V**

Part III   **The cloud as platform**

Data analytics	Streaming data	Machine learning	Research data portals
Spark & Hadoop	Kafka, Spark, Beam	Scikit-Learn, CNTK,	DMZs and DTNs, Globus
Public cloud Tools	Kinesis, Azure Events	Tensorflow, AWS ML	Science gateways

Managing data in the cloud
File systems
Object stores
Databases (SQL)
NoSQL and graphs
Warehouses
Part I   Globus file services

Computing in the cloud
Virtual machines
Containers – Docker
MapReduce – Yarn and Spark
HPC clusters in the cloud
Mesos, Swarm, Kubernetes
Part II   HTCondor

Part IV:
Building Your Own Cloud

We cover two topics in this fourth part of our book: how to build your open private cloud on your own hardware in your own institution, and how to build your own software as a service (SaaS) systems that run on public clouds.

A private cloud, as we explained in section 1.2 on page 3, is a cloud infrastructure operated for a single organization. A private cloud is generally taken to be a compute cluster that supports APIs similar to those provided by the Amazon EC2 and S3 services described in parts I and II. That is, it supports on-demand provisioning of virtual machine instances and storage buckets. Building a private cloud is thus a matter of deploying and running software that provide such APIs. Numerous software stacks have been developed for this purpose, of which the following are among the most frequently used.

- **OpenStack** `openstack.org` is an open source project that curates software for building private and public clouds. Its separate, individually accessible services can be deployed in various combinations, making it possible to customize an OpenStack deployment to meet specific private cloud demands.

- **OpenNebula** [202] `opennebula.org` is an open source cloud project that simplifies the process of deploying a private cloud. Because each data center is architected differently, by many architects, possibly over many years, developing a portable private cloud platform that can install in any data center configuration is difficult. OpenNebula addresses this challenge by providing a simple set of services that easily integrate with existing data center hardware, software, and administrative policies.

- **Eucalyptus** [212] is an open source project for building private and hybrid clouds that are API-compatible with Amazon AWS. It was designed to enable the creation of private clouds with a consistent API and functionality, regardless of how they are deployed; its API compatibility means that applications can be migrated without change between Eucalyptus and Amazon. Thus Eucalyptus is architected not as a set of separately developed services, but as an end-to-end integrated service ensemble.

- **Apache CloudStack** `cloudstack.apache.org` is an open source software project that bridges between the customizability and site-specific deployment characteristics of OpenStack and OpenNebula with the scale, reliability, and API portability of Eucalyptus. It supports its own API and also provides limited support for older versions of the AWS API.

- Microsoft's **Azure Stack** [32] is proprietary (i.e., non-open source) software that can be deployed within a data center, primarily to enable hybrid cloud operation with the Azure public cloud. It supports basic cloud functionality using the Azure APIs and includes extensive support for hybrid operation.

- **VMware Cloud Foundation** [50] offers a suite of proprietary virtualization technologies from which it is possible to build a private cloud. The Cloud Foundation product provides installation and deployment support for these technologies.

We describe two of these private cloud software stacks here: Eucalyptus, which has the dual merits of being particularly easy to deploy and of implementing Amazon APIs, in chapter 12; and the more complex but also more configurable, and perhaps for that reason more popular, OpenStack in chapter 13.

In chapter 14, we turn to the second topic of part IV, building your own software as a service. We explain that SaaS is both a technology and a business model [136]. As a technology, it features a single version of software, operated by a SaaS provider, that is consumed by many customers over the network. As a business model, it features lightweight pay-for-use or subscription-based compensation mechanisms that both minimize friction for consumers and enable SaaS providers to scale delivery with usage.

Together, these two concepts have proven remarkably successful in enterprise and consumer software, allowing previously expensive capabilities to be delivered to many more people at dramatically lower prices. We discuss the implications of SaaS for scientific software, review selected projects that deliver scientific software over the network, and present some examples of where SaaS proper is being applied in scientific settings. Space does not permit a comprehensive, step-by-step treatment of how to build a SaaS system, but we hope that the material here will whet your appetite for building your own software services.

Chapter 12

Building Your Own Cloud with Eucalyptus (with Rich Wolski)

"Cloud is about how you do computing, not where you do computing."

—Paul Maritz

Eucalyptus is an open source software infrastructure for implementing private clouds. It is structured as a set of cooperating web services that can be deployed in a large array of configurations, using standard, commodity data center hardware. It is packaged by using the Red Hat package manager [46] and tested extensively by Hewlett Packard, which serves as the primary curator of the open source project. Hewlett Packard also offers commercial extensions to Eucalyptus that enable it to incorporate non-commodity network and storage offerings from other vendors.

Eucalyptus has two key features that differentiate it from other private cloud IaaS platforms, such as OpenStack, CloudStack and OpenNebula. The first is that it is API compatible with Amazon. As a result, code, configuration scripts, VM images, and data can move between any Eucalyptus cloud and Amazon without modification. In particular, the large collection of freely available open source software (and all necessary configurations) on Amazon can be downloaded and run in a Eucalyptus cloud.

The second differentiating feature of Eucalyptus is that it is packaged to enable easy deployment onto the types of compute resources typically found in a data center (e.g., 10 gigabit Ethernet, commodity servers or blade chassis, storage area network devices, JBOD arrays); but once deployed, it operates in exactly the same way as every other Eucalyptus cloud—and, for that matter, in the same way as

Amazon. This design feature is useful for enterprises that wish to deploy a cloud to *use* but are not interested in developing their own new cloud technologies or creating a locally unique customized cloud.

The implementations of the cloud abstractions within Eucalyptus are designed end to end so that they work consistently across deployment architectures, are performance optimized, and are reliable. These features make Eucalyptus inexpensive to maintain as scalable data-center infrastructure, often requiring only a small fraction of the system administration support typically needed for other technologies and platforms. The disadvantage, however, is that Eucalyptus is more difficult to modify than other cloud platforms. Thus, enterprises wishing to develop their own proprietary cloud technologies often find it an inappropriate choice. In short, Eucalyptus is designed for those who wish to run a production private cloud, but not for those who wish to use it as a toolkit to build other technologies.

12.1   Implementing Cloud Infrastructure Abstractions

The cloud IaaS abstractions provide user access to virtualized resources that, once provisioned, behave in the same way as their data-center counterparts. Using cloud-hosted resources is different from using native bare-metal resources, however, because in the cloud resource provisioning and decommissioning are self-service and each resource is characterized by a service level agreement (SLA) or service level objective (SLO). That is, cloud users are expected to operate a provisioning API that allocates resources for their exclusive use (and similarly a decommissioning API when they have finished using the resources). Also, rather than acquiring access to a particular make and model of resource, users must consider the quality of service (described by an SLA or SLO) associated with that resource. The cloud is free to implement each resource request (using virtualization) with whatever resources are capable of meeting the SLA terms or the SLO.

For private clouds, self-service is implemented by using distributed, replicated, and tiered web services to automate resource provisioning. In Eucalyptus, requests are decomposed into subrequests that are routed to different services and handled asynchronously to improve scale and throughput. A request is ready for its user once all its subrequests complete successfully.

For example, to provision a VM, Eucalyptus decomposes the provisioning request (which specifies a VM type) into subrequests for

- a VM image containing either a Linux or Windows distribution;

- a fixed number of virtual CPUs and a fixed-size memory partition;

- one or more ephemeral disk partitions attached to the VM when it boots;

- a public and a private IP address for the VM;

- a MAC address to be used by the VM to make a DHCP request for other network information; and

- a set of firewall rules associated with the security group in which the VM is allocated.

Once the request is authenticated and any user-specific access control policies are applied, these resource-provisioning subrequests are initiated separately and handled asynchronously by one or more internal web services. The same decomposition approach is used for the other IaaS abstractions, such as disk volumes, objects in the object store, firewall rules, load balancers, and autoscaling groups.

Eucalyptus allows cloud administrators to determine the SLAs and SLOs that are to be supported via a site-specific deployment architecture that the administrator must define. Further, administrators are responsible for publishing the resulting SLAs and SLOs to their user community.

For example, in one installation at UCSB, the cloud is divided into two separate Availability Zones (AZs)—one containing newer, faster computing servers than the other. Thus a user requesting a specific VM type from one AZ gets a different speed processor (different cache size, different memory-bus speed, etc.) that the *same* VM type would deliver from the other AZ. It is up to the cloud administrator to publish what SLO a user should expect from each AZ, so that users of this cloud can reason about the VMs that they are provisioning.

Thus, in a private cloud, the deployment architecture determines the SLAs and SLOs that can be satisfied. This feature is often attractive in private data-center contexts, in which different organizations purchase hardware scoped to meet differing specific needs. Eucalyptus allows this hardware to be accessed in a uniform (and Amazon-compatible) manner, while also allowing the administrator to specify the SLAs and SLOs that the users can expect.

12.2 Deployment Planning

The Eucalyptus documentation describes the deployment planning process [16] and provides sample reference architectures that are suitable for different private cloud use cases. In this section, we describe some of the high-level trade-offs that typically arise when planning a Eucalyptus deployment.

12.2.1   Control Plane Deployment

The Eucalyptus control plane consists of the following cooperating web services, which communicate via authenticated messages:

- Cloud Controller (CLC), which manages the internal object request lifecycle and cloud bootstrapping

- Cluster Controller (CC), which manages a cluster or partition of compute nodes

- Storage Controller (SC), which implements the network-attached block-level storage abstraction (e.g., Amazon EBS)

- Walrus, which implements the cloud object store (e.g., Amazon S3)

- Node Controller (NC), which actuates VMs on a compute node

- User Facing Services (UFS) component, which fields and routes all user-requests to the appropriate internal services

- Eucalyptus Management Console, which implements the graphical user console and cloud administration console

In addition, the CC uses a separate component called Eucanetd to handle the various networking modes that Eucalyptus supports. In some networking modes, Eucanetd must be co-located with the CC; in others, it is co-located with the NC.

These services can be deployed in a variety of ways: all together on the same node (as we describe in section 12.3 on page 267), on separate nodes (one service per node), or in any combination. Separating the services so that they run on different nodes improves availability. Eucalyptus continues to function (possibly with a degradation of service) when one or more of its internal services become unavailable. Thus, separating services increases the chance that a node failure can be masked by the cloud. On the other hand, co-located services involve less installation time and require less hardware dedicated to the control plane.

The cloud administrator must also decide how many AZs to configure. Eucalyptus treats each AZ as a separate cluster. That is, a compute node hosting a VM can be in only one AZ. Each AZ requires its own CC service and SC service. Neither the CC nor SC services for two different AZs can be on the same host. However, the CC-SC pair for each of two AZs can be co-located if desired.

Each node that hosts a VM must run an NC, which acts as an agent for Eucalyptus. VM provisioning requests ultimately translate into commands to the NC, causing it to assemble and boot a VM on the local hypervisor and to attach the new VM to the Eucalyptus-provisioned network.

Beyond simple TCP connectivity, there are some additional connectivity requirements between the components when they are hosted on separate machines. The Eucalyptus installation documentation provides specifics [18].

12.2.2   Networking

Perhaps the most complex set of choices to make when planning a Eucalyptus deployment relate to the provisioning of virtual networks. To implement cloud connectivity and network security, Eucalyptus must be able to set up and tear down virtual networks within the data center. Typically, the network architecture in each data center is unique. Furthermore, the network infrastructure is often the vehicle for implementing security policies, which often prescribe a limited number of feasible network control options for a particular deployment. Eucalyptus supports several networking modes [19] (including Software Defined Networking), allowing the administrator to decide on the best approach for a specific data center.

The two most popular modes are `MANAGED-NOVLAN`, which uses the node that is hosting the CC service as a "soft" Layer 3 IP router that the CC can program dynamically, and `EDGE`, in which each node hosting a VM also acts as a router to implement network virtualization. The former has the advantage of being simple to configure and troubleshoot. However, it does not implement full Layer 2 network isolation, and thus a VM can snoop Ethernet network packets from the network to which the node hosting it is attached. Also, if the node hosting the CC goes down, VMs lose their external connectivity until it is restored. In contrast, `EDGE` mode implements both Layer 2 and Layer 3 network isolation and does not route network traffic through the node hosting the CC. However, changes to cloud network abstractions, such as those employed by security groups, take longer to propagate, due to the use of eventual consistency mechanisms.

12.2.3   Storage

A number of deployment options are available for the various cloud storage abstractions supported by Eucalyptus. For the object storage, the cloud can use a Linux file system on the node that is hosting the Walrus service: either RIAK CS [43] or Ceph `ceph.com`; each has its own installation and maintenance complexity, failure resilience properties, and performance profile. Similarly, for network-attached volume storage, Eucalyptus can use the local Linux file system on the node hosting the SC, Ceph via its RADOS interface [42], or one of several storage area network offerings from various vendors.

In general, the deployments that use the local Linux file system are simple to configure and maintain and are relatively performant. They do not replicate data, however, so an unmitigated storage failure can cause data loss. Some cloud administrators use the software RAID capability of Linux [28] to implement the file system that backs Walrus and the SC in this deployment configuration. When data loss is a strong concern, however, one of the other replicating storage technologies is usually less complex to maintain.

12.2.4 Compute Servers

On each compute server hosting an NC, the cloud administrator gets to specify VM sizing and virtual CPU speed. When a VM is initiated, it is allocated some number of cores, a fixed memory size, and ephemeral disk storage that appears as separate attached disk devices in the VM. These requirements are carried in the VM type, and the cloud administrator determines what VM types a specific cloud supports. Eucalyptus uses the core count, allocated disk space, and available memory to determine the maximum size VM type that can be hosted on each node. Additionally, Eucalyptus can use hypervisor multiplexing to overprovision the servers in terms of core counts, in which case the cores are time sliced.

These configuration operations mean that when planning a deployment, the cloud administrator must typically determine how much local disk storage (and from what disk partition) is to be used for VM ephemeral storage; whether to enable hardware hyperthreading (if it is available); and the degree to which hypervisor timeslicing of cores should be employed. Each of these parameters controls, in some measure, the SLO that a hosted VM can achieve.

12.2.5 Identity Management

Eucalyptus supports the same role-based identity management and request authentication mechanisms and APIs as Amazon does. This feature is particularly important, both for security reasons (Amazon is generally considered secure) and for API compatibility reasons. However, the deployment dictates how user credentials and role definitions are managed. In particular, it can operate as a standalone cloud, in which the cloud administrator is responsible for credential administration (e.g., credential distribution, revocation, role-definition policies), or it can be integrated with the data center's existing Active Directory or LDAP installation.

12.3   Single-cluster Eucalyptus Cloud

We illustrate the process of deploying a Eucalyptus private cloud in a single computational cluster. One node (machine) in the cluster acts as the **head node** that hosts all of the web services that compose the Eucalyptus control plane. In this configuration, all nodes except the head node host VMs. We call the nodes that host VMs **worker nodes**. Cloud requests (made via HTTPS or the Management Console) are fielded by the various services on the head node and, once authenticated and determined to be feasible, are forwarded to one or more NCs running on worker nodes for actuation. Similarly, when a request is terminated, the head node sends notice of the termination to all NCs that must deallocate resources associated with the request. The request is fully terminated when all NCs report successful deallocation.

This configuration is useful for a supported production deployment in many academic or research settings where a moderate-sized user population (e.g., an instructional class, research group, or development team) shares a cluster, also of moderate size (tens to hundreds of nodes). Note that the scalability of this configuration is typically determined by the number of nodes and not the total number of cores (separate CPUs) that each node comprises. Also, from a reliability perspective, all VMs remain active and network reachable in the event the head node fails or goes off line. No new cloud requests can be serviced while the head node is down and some storage abstractions cease to function; but VM activity, network connectivity, and access to ephemeral storage (which is local to each VM) are not interrupted with a head node failure. Further, functionality is completely restored when the head node is restored to functionality. Thus, this configuration, which is relatively simple to deploy and is portable to a wide variety of hardware configurations, is capable of long-duration VM hosting.

A single-cluster configuration typically requires little data-center support: commodity servers connected to a publicly routable subnet are sufficient to support a cloud. The cloud administration effort required for such an installation is also low: once the cloud is deployed, the cloud administrator is responsible for issuing user credentials, managing resource quotas, and setting instance type configurations. In an academic setting, this burden is usually budgeted as a small fraction of a local system administrator's available time.

12.3.1 Hardware Configuration

We consider for the following example installation a hardware configuration comprising four x86_64 servers, each with a single gigabit Ethernet interface attached to a publicly visible IPv4 subnet. Each server has four cores, 8 GB of memory, and 1 TB of attached storage. Eucalyptus is designed to function properly on a wide variety of server configurations. For the head node, generally 8 GB is necessary, but the worker nodes can have almost any configuration. However, the core counts, memory sizes, and available local storage on the worker nodes determine the maximum size for any instance type that an administrator can configure for the cloud.

12.3.2 Deployment

All services running on the head node (except the CC and the Management Console) share a single Java Virtual Machine (JVM). The available disk space on the head node is split between Walrus and the SC, and no software RAID is present. Here we use the **EDGE** networking mode, which requires each worker node to run both an NC service and a Eucanetd process.

In addition, the cloud requires a pool of available IP addresses from the same subnet to which the head node and worker nodes are attached, for assignment to hosted VMs. This configuration allows all VMs within the cloud to be reachable as if they were hosts on the same publicly visible subnet as the head and worker nodes. Further, these IP addresses cannot be in use by other hosts on the subnet: they must be available, but unassigned, within the subnet address space.

In this example, we assume a publicly routable subnet of 128.111.49.0/24, with 255 available IP addresses, and with 128.111.49.1 as the gateway for the subnet. We also assume that the nodes have been assigned 128.111.49.10 through 128.111.49.13 by the local network administrator but that all addresses between 128.111.49.14 and 128.111.49.254 on the subnet are available to the cloud for VMs. We further assume that the head node has the public IP address 128.111.49.10.

In addition, Eucalyptus assigns each VM both an internal private IP address and an externally routable public IP address (as does Amazon EC2). Thus, the cloud needs a private IP address range to use for VM private addresses. Because this network is private to the cloud, it can be any private network address range. We use 10.1.0.0/16 as the private address space for the cloud in this example. To allow each worker node to implement a firewall and router for the VMs it hosts, it needs a network address on this private subnet as well. We assume that the worker nodes get `10.1.0.2` through `10.1.0.4` and that the others are available for VMs. Note that the head node does not need an address from the private address range.

To keep the integration points between the cloud and the existing data center administration to a minimum, we assume that the cloud administrator manages user accounts either via scripts or via the Eucalyptus Management Console.

Figure 12.1 shows the single-cluster deployment that we use in this chapter as an example Eucalyptus deployment.

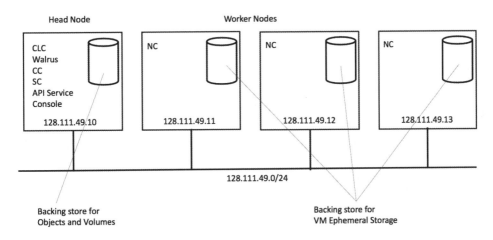

Figure 12.1: Example single-cluster Eucalyptus deployment, showing the one head node running management, storage, and networking services and three worker nodes each running NC software.

Eucalyptus also assigns internal and externally resolvable DNS names to each VM that it hosts. To do so, it requires a cloud-local subdomain for the private cloud that it is to manage. In the example, we use the subdomain name `testcloud.ucsb.edu`. The Domain Name System (DNS) server that the nodes use for DNS service must be configured to forward DNS name requests for externally resolvable instance names to the head node on port 53 (the standard DNS port).

12.3.3 Software Dependencies and Configuration

The current version of Eucalyptus requires that nodes run the latest version of either Red Hat Enterprise Linux (RHEL) version 7 or the Community-supported version (CentOS) version 7. All nodes must run the Network Time Protocol `ntp` [86] so that the Secure Software Layer (SSL) can protect against request replay attacks [30].

Eucalyptus, in EDGE mode, requires a number of Layer 3 IP ports to be opened so that the internal services can communicate with each other, with software dependencies within Linux, and with users and administrators. Information about these ports and their functions is available online [17]. Ports can be opened by the root user individually [10]—or, if the subnet that the nodes are using is protected by another firewall, the root user can make all ports accessible by executing the following command. (Note that this command opens *all* ports, so care must be taken to ensure that the system is otherwise secure.)

```
systemctl stop firewalld
```

For EDGE mode, the worker nodes must attach a Linux network bridge to the network interface [53]. Below, we show the bridge configuration steps for the first worker node, which has the IP address 128.111.49.11. You must first install the bridge utilities software with this command:

```
yum -y install bridge-utils
```

In the directory /etc/sysconfig/network-scripts, create the file ifcfg-br0, and place in that file the following text:

```
DEVICE=br0
TYPE=Bridge
BOOTPROTO=static
ONBOOT=yes
DELAY=0
GATEWAY=128.111.49.1
IPADDR=128.111.49.11
NETMASK=255.255.255.0
BROADCAST=128.111.49.255
IPADDR1=10.1.0.2
NETMASK1=255.255.0.0
BROADCAST1=10.1.255.255
NM_CONTROLLED=no
DNS1=128.111.49.2
```

Note that the bridge needs both the public IP address for the node and an address on the private subnet that is to be used by VMs. In this example, we have chosen 10.1.0.2. Also, you need to specify the IP address of the DNS service that this node should use: in this case, 128.111.49.2.

Next, run the command `ip addr show` to determine the system-chosen name for the Ethernet interface. RHEL CentOS 7 uses a dynamic naming scheme for network addresses. To determine what name it has chosen for the Ethernet interface, look for the IP address attached to a device that is marked as UP.

To bridge this device, find in the directory `/etc/sysconfig/network-scripts` the file that begins with the string `ifcfg-` and ends with the device name. For example, if Linux has named the Ethernet device `enp3s0`, then you want the file named `ifcfg-enp3s0` in that directory. You need the Ethernet MAC address for this device as well. To get it, run the command `ip link` and look for the `link/ether` field for the device. The MAC address is the colon-separated address, which in our example is `00:19:b9:17:91:73`.

```
DEVICE=enp3s0
# Change   hardware address to that used by your NIC
HWADDR=00:19:b9:17:91:73
ONBOOT=yes
BRIDGE=br0
NM_CONTROLLED=no
```

The network must be restarted in order for this change to take effect. To do so, issue the following command as the root user and then check with `ip addr show` that the bridge now carries the public and private IP addresses.

```
systemctl restart network
```

12.3.4   Installation

You must first temporarily disable the firewall on each node by running the following command. (This command and all other commands in this section need to be run as the root user.)

```
systemctl stop firewalld.service
```

Note that the firewall will be re-enabled (if it was already enabled) when the system reboots. If you reboot the system for any reason before the installation is complete, you must repeat this step. After the installation is complete, you can re-enable the firewall service.

Next, install the Eucalyptus release packages `downloads.eucalyptus.com`. At the time of writing, the latest release of Eucalyptus is version 4.3 and of Euca2ools is version 3.4. Thus you run the following commands:

```
PREFIX="http://downloads.eucalyptus.com/software"
yum -y install ${PREFIX}/eucalyptus/4.3/rhel/7/x86_64/eucalyptus-release-4.3-1.el7.noarch.rpm
yum -y install ${PREFIX}/euca2ools/3.4/rhel/7/x86_64/euca2ools-release-3.4-1.el7.noarch.rpm
yum -y install http://dl.fedoraproject.org/pub/epel/epel-release-latest-7.noarch.rpm
```

You then install the node controllers on the worker nodes. Each worker node needs to run a Eucalyptus NC. To install the NC code and the network virtualization daemon (Eucanetd), run the following command sequence on each worker node. The last two `virsh` commands remove the default libvirt network so that Eucanetd can start a DHCP server.

```
yum -y install eucalyptus-node
yum -y install eucanetd
systemctl start libvirtd.service
virsh net-destroy default
virsh net-autostart default --disable
```

You need to check that KVM did not have the virtual network busy, by running `ls -l /dev/kvm`. If the output does not show the KVM device as being accessible, as follows, then reboot the machine to clear the libvirt lock on the device.

```
crw-rw-rw-+ 1 root kvm 10, 232 Jan 24 13:03 /dev/kvm
```

The final step installs the cloud on the head node. On the head node, run the following commands to install the Eucalyptus control plane services plus an imaging service (that runs as a VM) for converting different image formats automatically within the cloud.

```
yum -y install eucalyptus-cloud
yum -y install eucalyptus-cluster
yum -y install eucalyptus-sc
yum -y install eucalyptus-walrus
yum -y install eucalyptus-service-image
yum -y install eucaconsole
```

12.3.5 Head Node Configuration

You must configure the Security-Enhanced Linux (SELinux) kernel security module to allow Eucalyptus to function as a trusted service. On the head node, run the following commands as the root user:

```
setsebool -P eucalyptus_storage_controller 1
setsebool -P httpd_can_network_connect 1
```

Next, you configure `EDGE` networking on the head node. You do this by editing the file /etc/eucalyptus/eucalyptus.conf and setting `VNET_MODE="EDGE"`. Note that the # character serves as a comment. Also on the head node, you need to create a JSON file that contains the network topology that the cloud is to use for each VM that it hosts. The file format is documented in the Eucalyptus EDGE Network Configuration manual [15]. For the example hardware configuration described in

```json
{
    "InstanceDnsDomain": "eucalyptus.internal",
    "InstanceDnsServers": ["128.111.49.2"],
    "MacPrefix": "d0:0d",
    "PublicIps": [
        "128.111.49.14-128.111.49.254"
    ],
    "Subnets": [
    ],
    "Clusters": [
        {
            "Name": "az1",
            "Subnet": {
                "Name": "10.1.0.0",
                "Subnet": "10.1.0.0",
                "Netmask": "255.255.0.0",
                "Gateway": "10.1.0.1"
            },
            "PrivateIps": [
                "10.1.0.5-10.1.0.254"
            ]
        }
    ]
}
```

Figure 12.2: Example JSON configuration file for a Eucalyptus virtual network.

section 12.3.1 on page 268, you create the file /etc/eucalyptus/network.json with the contents shown in figure 12.2.

Note that the public and private IP addresses used by the head node and worker nodes are *not* in the list of addresses. Eucalyptus uses addresses from the list in this file for VMs; thus, these addresses must not conflict with the addresses used by the nodes hosting the cloud. Note also that the Cluster has a name (az1 in the example). This is the name of the availability zone (you can set it to any string you like). You need this name when you go to register the availability zone with the cloud.

In this example, the head node uses the file system as backing store for Walrus and the SC. By default, the directories /var/lib/eucalyptus/bukkits are used by Walrus and /var/lib/eucalyptus/volumes by the SC as top-level directories for the backing store files. A typical CentOS installation on a machine with a single disk creates two separate partitions for the root file system and for home directories and puts the bulk of the available disk space in the home directory partition. If you leave the `bukkits` and `volumes` directories on the root partition, it may run out of disk space.

Eucalyptus follows symbolic links for these two directories. If the root partition is small compared with your home partition, then create symbolic links by running the following commands as the root user. (Alternatively, you can reconfigure your root partition to have more space and avoid the use of these symbolic links.)

```
rm -rf /var/lib/eucalyptus/bukkits
rm -rf /var/lib/eucalyptus/volumes
mkdir -p /home/bukkits
mkdir -p /home/volumes
chown eucalyptus:eucalyptus /home/bukkits
chown eucalyptus:eucalyptus /home/volumes
ln -s /home/bukkits /var/lib/eucalyptus/bukkits
ln -s /home/volumes /var/lib/eucalyptus/volumes
chmod 770 /home/bukkits
chmod 770 /home/volumes
```

12.3.6 Worker Node Configuration

On each worker node, edit the file `/etc/eucalyptus/eucalyptus.conf`, and set the following keyword parameters

```
VNET_MODE="EDGE"
VNET_PRIVINTERFACE="br0"
VNET_PUBINTERFACE="br0"
VNET_BRIDGE="br0"
VNET_DHCPDAEMON="/usr/sbin/dhcpd"
```

Each NC maintains a cache of VM images as well as a backing store for running instances in the local file system. The path to the top-level directory for these storage requirements is given by the `INSTANCE_PATH` key-word parameter in the file `/etc/eucalyptus/eucalyptus.conf`, and its default value is `/var/lib/eucalyptus/instances`. To move this to the home disk partition (see the discussion of disk space in the preceding subsection), run the following commands as the root user on each worker node:

```
mkdir -p /home/instances
chown eucalyptus:eucalyptus /home/instances
chmod 771 /home/instances
```

Next, edit the file /etc/eucalyptus/eucalyptus.conf, and set the key-word parameter INSTANCE_PATH to be /home/instances. Installation is complete.

12.3.7 Bootstrapping

Eucalyptus needs to go through a one-time bootstrapping step after a clean install. Note, however, that Eucalyptus also supports upgrades between versions; the bootstrapping process described here is needed only after the packages are installed for the first time. The bootstrapping process uses a set of command-line tools that are installed on the head node, as shown in figure 12.3 on the next page. Some of these tools are specific to head node operation, while others are part of the standard Amazon command line tools that are part of the **euca2ools** [14] command line interface. In what follows, you run all the commands from the Linux shell as a root user for the node.

12.3.7.1 Registering the Eucalyptus Services

The next step is to register the various service components with each other. Registration requires that the head node use the Linux command **rsync** to transfer configuration state. As such, it is easiest if the head node can use **rsync** without a passphrase, both to **rsync** with each worker node and with itself [170]. Otherwise, each registration step prompts the user to enter the root password either on the head node or on the specific worker node that is being registered, possibly several times for each. Passphrase-less **rsync** can subsequently be disabled once registration is complete.

Eucalyptus works best if it uses the public IP addresses rather than the DNS names of the nodes for registration. Also, you need the name of the AZ specified in the network topology JSON file (**az1** in this example).

Once all services are running (the registration step cannot take place when the services are down or not yet ready), run this command on the head node to generate a set of bootstrapping credentials.

```
eval clcadmin-assume-system-credentials
```

This command sets shell environment variables containing a temporary set of credentials that allow subsequent commands to stitch the cloud services together securely. Thus, you must use this shell for the remaining registration steps.

To register the user-facing services, determine the public IP address of the head node, and choose human-readable names for the services. In this example,

```
# First run this command on the head node.
clcadmin-initialize-cloud

# Run these commands to have CentOS 7 bootstrapper restart cloud
# automatically when head node reboots. (The tgtd service is needed
# for the SC to be able to export volumes for VMs.)
systemctl enable eucalyptus-cloud.service
systemctl enable eucalyptus-cluster.service
systemctl enable tgtd.service
systemctl enable eucaconsole.service

# Start the control plane services on the head node
systemctl start eucalyptus-cloud.service
systemctl start eucalyptus-cluster.service
systemctl start tgtd.service
systemctl start eucaconsole.service

# (Optionally) enable the node controller to restart after a reboot
systemctl enable eucalyptus-node.service
systemctl enable eucanetd.service

# Start the node controller
systemctl start eucalyptus-node.service
systemctl start eucanetd.service

# Check that all components are running by running:
netstat -plnt
# and verifying that there are processes listening on ports 8773 and
# 8774 on the head node and 8775 on the worker nodes.
# (Note that it may take a few minutes for services to be visible.)
```

Figure 12.3: Commands used to bootstrap a Eucalyptus cloud.

the public IP address for the head node is `128.111.49.10`, and we use the name `ufs_49.10` as the service name. To register the example user-facing services, you run the following command on the head node.

```
euserv-register-service -t user-api -h 128.111.49.10 ufs_49.10
```

Next, register the backend service for the Walrus object store. Again, using the public IP address for the head node and a service name, the registration command for the example is as follows.

```
euserv-register-service -t walrusbackend -h 128.111.49.10 walrus_49.10
```

The registration procedure for the CC and the SC is similar, but it requires the AZ name from the network topology JSON file for the `-z` parameter. The registration commands for the example are as follows. The third command installs security keys in the appropriate place in the head node file system.

```
euserv-register-service -t cluster -h 128.111.49.10 -z az1 cc_49.10
euserv-register-service -t storage -h 128.111.49.10 -z az1 sc_49.10
clcadmin-copy-keys -z az1 128.111.49.10
```

To register the NC services running on each worker node, you must run the node registration commands on the head node giving the IP address for each worker node. In the example, these commands are as follows.

```
clusteradmin-register-nodes 128.111.49.11 128.111.49.12 128.111.49.13
clusteradmin-copy-keys 128.111.49.11 128.111.49.12 128.111.49.13
```

12.3.7.2 Runtime Bootstrap Configuration

With the services running and securely registered, the last bootstrapping step is to configure the runtime system.

To configure DNS name resolution for cloud instances, you need the name of the subdomain that is to be forwarded by the site DNS service to the head node. In this example, we use the name `testcloud.ucsb.edu`. Thus, on the head node you run the following commands:

```
euctl system.dns.dnsdomain=testcloud.ucsb.edu
euctl bootstrap.webservices.use_instance_dns=true
```

The head node acts as the authoritative DNS service for names in the associated subdomain: `testcloud.ucsb.edu` in our example. To test the linkage, run this command on the head node:

```
host compute.testcloud.ucsb.edu
```

It should resolve to the head node's public IP address. If it does not, check the configuration of the site DNS to ensure that it is forwarding name requests for the cloud subdomain to the head node.

Next, create permanent administrative credentials for the cloud. These credentials allow the cloud administrator full access to all resources (i.e., they are the super-user credentials for the cloud). For security purposes, the SSL used internally uses DNS resolution as part of its antispoofing authentication tests. Thus you need to use the cloud subdomain specified in the previous command when generating the administrator credentials. For the example, you run the following commands on the head node.

```
cd /root
mkdir -p .euca
euare-usercreate -wld testcloud.ucsb.edu adminuser >\
        /root/.euca/adminuser.ini
```

You also need to tell the local command line tools that you wish to contact this cloud (as opposed to other clouds or Amazon itself) by setting the region to the cloud's local subdomain. For the example cloud, you run the following commands in the shell where you were using the temporary credentials.

```
eval euare-releaserole
export AWS_DEFAULT_REGION=testcloud.ucsb.edu
```

If you want the root user always to contact the cloud running on the head node, add these commands to the file /root/.bashrc; they are then set when the root user logs in.

The next step is to upload the network topology JSON file to the cloud. With the permanent administrative credentials installed as described previously, run the following command.

```
euctl cloud.network.network_configuration=@/etc/eucalyptus/network.json
```

Next, the storage options for the cloud must be configured. In this example, we are using the local file system on the head node for both object storage and volume storage. For the volume storage configuration, you need to specify the name of the AZ (az1 in this example). You use the following commands to enable this storage configuration.

```
euctl objectstorage.providerclient=walrus
euctl az1.storage.blockstoragemanager=overlay
```

To enable the imaging service, run the following command using the local cloud subdomain as the region.

```
esi-install-image --region testcloud.ucsb.edu --install-default
```

Eucalyptus then installs a VM that can import raw disk images for use as volume-backed instances with this service.

12.3.7.3 Quick Health and Status Checks

At this point, your cloud should be up and functional. To ensure that it is working, run the following status command.

```
euserv-describe-services
```

All services should report in the state **enabled**. To determine available instance capacities, execute the following command with administrator credentials.

```
euca-describe-availability-zones verbose
```

If all NCs are properly registered, the sum of their capacities should be displayed. In this example, each worker node supports four cores. Thus the cloud should be able to run 12 `m1.small` instance types when all three NCs are registered and no other VMs are running.

12.3.8 Image Installation

Eucalyptus maintains a repository of network accessible curated images that can be installed automatically. To install from the image repository, as the root user with the administrator credentials enabled, run the following commands.

```
yum install -y xz
bash <(curl -Ls eucalyptus.com/install-emis)
```

The installation script then prompts you for the images to install. This script also checks to make sure that all of the dependencies needed to install images are present. If not, the script prompts you to ask whether it should install any that are missing using the `yum` utility.

Note that the image installation is using the cloud administrator credentials. As a result, the image is accessible only by the cloud administrator. To make it available to all users, run the following command. Note the image identifier in the output that begins with the string `emi-`.

```
euca-describe-images -a
```

You will also see in the output another installed image with an `emi-` identifier: the image that hosts the imaging service. Choose the identifier for the image that you just installed.

Then, run the `euca-modify-image-attribute` with the `emi-` identifier, set the `-a` flag to `all`, and add the `-l` flag. For example, if the image installation installed `emi-1e78481f`, then run the following command to set the launch permissions so that *all* accounts may launch an instance from the image.

```
euca-modify-image-attribute -a all emi-1e78481f -l
```

12.3.9 User Credentials

The cloud administrator can create accounts for users other than the administrative user. Each account has its own administrative user that can create others users in the account. Unlike the cloud administrator, however, these account administrators do not have access to resources outside of their specific accounts.

To create a user account, you need a unique account name. For example, to create an account for `user1`, run the following command.

```
euare-accountcreate -wl user1 -d testcloud.ucsb.edu > user1.ini
```

This command outputs a credentials file that this user can install in the user's `.euca` directory for use with the cloud. The user must also set the `AWS_DEFAULT_REGION` environment variable to the name of the local cloud DNS subdomain (`testcloud.ucsb.edu` in this example). These credentials allow the user to access the cloud via the command line interface.

To enable the user to access the cloud from the Eucalyptus Management console, run the following command, with `initialpassword` being the password that you use to gain initial access (and should change).

```
euare-useraddloginprofile --as-account user1 -u admin -p initialpassword
```

With this password, you can point a web browser to the head node and attempt to log in. In the test example, you would use the URL `https://128.111.49.10`

to contact the Management console. The certificate is self-signed, so most browsers will ask you to confirm that they wish to make a security exception. At the login screen, you then need to enter `user1` as the account name, `admin` as the user in that account, and the password. Once logged in, you can change the password. Note that this password is only for the Management Console. The command line tools use Amazon-style credentials that are embedded in the `.ini` file generated when the account was created.

12.4   Summary

We have described how to build a private cloud using Eucalyptus. We first explained how private clouds implement the cloud abstractions. Eucalyptus supports API compatibility with Amazon and implements the same cloud abstractions so that workloads and data can move seamlessly between any Eucalyptus clouds, regardless of deployment architecture, and also between Eucalyptus and Amazon. We also discussed the role that deployment architecture has on SLAs and SLOs in a private cloud. We concluded the chapter with a step-by-step description of how to deploy a production-ready Eucalyptus private cloud using commodity servers connected to a local-area network. The deployment steps comprise cloud software installation and configuration, secure bootstrapping of the cloud, and initial administrative actions necessary to make the cloud available for users.

12.5   Resources

The Eucalyptus website `www.eucalyptus.com` provides access to a wide range of documentation and examples, as well as the Eucalyptus code.

Chapter 13

Building Your Own Cloud with OpenStack (with Stig Telfer)

*"If we as a society do not understand "the cloud," in all its aspects –
what data it holds, how it works, what the bargains are we make as we
engage with it, we'll all be the poorer for it, I believe."*

—John Battelle

OpenStack is an open source cloud operating system used by a broad and growing global community [54]. The Openstack software has a twice-yearly release cycle, with versions named in an incrementing alphabetical sequence. The latest release at the time of writing, OpenStack Newton [38], includes code contributions from 2,581 developers and 309 organizations. OpenStack's development model embodies the state of the art in distributed, open source software development.

While OpenStack's origins are in the orchestration of virtual machines, the project has diversified to become a versatile coordinator of virtualization, containerization and bare metal compute. OpenStack now provides a unified foundation for the management of many forms of storage, network, and compute resources. User surveys have identified four dominant use cases [80]:

- Enterprise private cloud

- Public cloud

- Telecom and network functions virtualization

- Research and big data, including high-performance computing

OpenStack's popularity as a choice for research computing infrastructure management is underlined by its use in academic clouds in the U.S., such as **Chameleon** `chameleoncloud.org`, **Bridges** [51], and **Jetstream** `jetstream-cloud.org`, and in international projects such as NeCTAR `cloud.nectar.org.au` in Australia and at CERN [37] in Europe. The scientific computing use cases of OpenStack are served by a dedicated area `openstack.org/science` within the OpenStack website.

13.1 OpenStack Core Services

OpenStack control planes are formed from a number of intercommunicating (mostly) stateless services and a set of stateful data stores. OpenStack services are described in depth in an online resource called the OpenStack project navigator `openstack.org/software`. Table 13.1 lists core components.

The cloud infrastructure ecosystem is evolving rapidly. The OpenStack strategy for adapting to a rapid pace of innovation is referred to as the "Big Tent." New projects, often experimental or exploratory, can easily be created and are encouraged to progress toward the OpenStack conventions for project governance as their functionality develops. Competing or conflicting projects are supported, to enable constructive competition within the ecosystem.

13.2 HPC in an OpenStack Environment

OpenStack can be configured variously to orchestrate compute resources in three different ways: via virtualization, containerization, or bare metal. (Recall that we

Table 13.1: Six core components likely to be found in any OpenStack deployment.

Name	Function	Description
Keystone	Identity	Provides authentication and authorization for all OpenStack services
Nova	Compute	Provides virtual servers on demand
Neutron	Networking	Provides "network connectivity as a service" between interface devices
Glance	VM images	Supports discovery, registration, and retrieval of VM images
Cinder	Block storage	Provides persistent block storage to guest VMs
Swift	Object storage	Object storage interface

described the difference between virtual machines and containers in chapter 4; in a bare metal deployment, software is run directly on the underlying hardware.) Outside HPC, OpenStack is most frequently used as an orchestrator of virtualization, in order to realize the full flexibility and advantages of software-defined infrastructure. In contrast, bare metal deployments are most common in HPC settings. While HPC workloads can be run on OpenStack systems configured in any one of the three forms, many administrators choose to trade off flexibility for reduced runtime overhead. However, because of rapid technological evolution at all levels, this trade-off is a continually shifting balance.

One emerging trend is to use specialized hardware and virtualization optimizations to support virtualized HPC workloads in ways that deliver the flexibility of software-defined infrastructure but avoid associated performance overheads. Hardware technologies such as Single-Root I/O Virtualization (SR-IOV), described in a later section, integrate with virtualized OpenStack compute to deliver HPC networking with minimized overhead. Virtualization optimizations such as CPU pinning and non-uniform memory access (NUMA) passthrough, both also described later, enable hardware-aware scheduling and placement optimizations. This ability to reconfigure cloud infrastructure via programming offers advantages throughout the software development life cycle. You can, for example, develop and test an application or workload on a standard OpenStack system and then apply hardware and virtualization optimizations.

We focus in this chapter on OpenStack deployments that use virtualization. Information available online describes implementation of containerized [12] and bare metal [9] use cases.

13.3 Considerations for Scientific Workloads

OpenStack's configuration and services can be adapted to support a range of requirements particular to scientific workloads. We describe some examples here.

13.3.1 Network-intensive Ingest or Egress

An OpenStack workload may involve the ingest and/or egress of large volumes of data involving external sources: for example, data ingested from a scientific instrument or a public dataset not hosted within the cloud. In these circumstances, the external network bandwidth of compute instances can become a bottleneck.

A typical OpenStack configuration may deploy **software gateway routers** for networking between compute instances and the external world. However, while such software gateway routers implement the rich feature set of software-defined networking, they struggle to deliver high bandwidth and low latencies. When operating in extremis, software switches are observed to discard packets instead of exerting back pressure. The following alternative configurations can be used to improve performance for external network connectivity.

- **Provider networks.** A provider network is a pre-existing network in the data center. It is not created or controlled by OpenStack Neutron, the OpenStack networking controller, but Neutron can be made aware of it and connect compute instances with it. This approach bypasses OpenStack-controlled routing and gateways.

- **Router gateways in silicon.** Some switch vendors are able to offload software-defined networking (SDN) capabilities into switch port configurations, enabling OpenStack-defined Layer-3 Internet protocol routing operations to be performed at full speed in the switch ports. Similarly, some network interface cards (NICs) support hardware offloading of large portions of SDN, greatly reducing load on control plane network nodes. Router gateways in silicon do not support rich networking features such as network address translation (NAT, required for supporting floating IP addresses), although these features may not be necessary for private cloud use cases.

- **Distributed virtual routers.** Distributing network node functions over components within each hypervisor produces a scalable external router implementation. (The **hypervisor** is the software component responsible for creating and running virtual machines. Typically there is one hypervisor instance per compute node.) One drawback of this approach is the increased hypervisor CPU overhead for networking. Furthermore, the approach raises security concerns by making every hypervisor externally reachable.

13.3.2 Tightly Coupled Compute

In a generic OpenStack configuration, networking for compute instances passes through one or more software virtual switches in the hypervisor, providing flexibility in configuration but leading to higher latency and reduced bandwidth for applications due to the additional data copies and context switches. Such software switches can also introduce higher levels of jitter and packet loss. The performance of tightly coupled application workloads, such as some bulk synchronous parallel applications, can be strongly influenced by communication latencies between instances. Consequently, virtualized networking can have an adverse impact on the performance of tightly coupled workloads.

The overheads introduced by virtualized networking can be bypassed through use of **Single-Root I/O Virtualization** (SR-IOV), although this feature is not supported by all NICs. This PCI hardware capability specifies how a PCI device can be shared, through creation of a number of shadow devices referred to as **virtual functions**. A hypervisor that supports SR-IOV enables the passthrough of a virtual function device into a compute instance, giving the virtual instance direct access to the underlying physical network device's hardware interface. The resulting direct path from compute instances to physical networks circumvents the software-defined networking implementation by the hypervisor. While this approach delivers high levels of performance, it also bypasses the security group firewall rules that OpenStack applies to an instance. Consequently, SR-IOV networking should only be used on internal (trusted) networks and should be configured in conjunction with conventional network configurations for managing connectivity with untrusted networks.

The performance of latency-sensitive workloads can be improved further through smarter process scheduling. Pinning virtual processor cores to physical cores improves cache locality. Memory access performance is improved by leveraging affinity between physical processor cores and memory regions. We briefly explain these concepts. Modern system architectures tend to incorporate multiple processors, each with an integrated memory controller and with memory directly attached to each processor. A single-memory system is constructed with coherent access to all memory from all CPU cores, with hardware buses between processors to ensure consistency. A consequence of this design is that memory and CPUs are unevenly coupled: what is referred to as non-uniform memory access. Making the virtual compute instances aware of the NUMA topology of the physical host enables better scheduling and placement decisions by the guest kernel. The compute hypervisors and OpenStack services can be configured to enable these optimizations.

13.3.3 Hierarchical Storage and Parallel File Systems

A workload may require high-performance coupling with a data source rather than between compute hosts. OpenStack can support storage services of multiple types, including types suitable for different tiers in a storage hierarchy, concurrently. However, it itself does not have a native implementation of hierarchical storage management. The HPC data-movement protocol **iSCSI Extensions for RDMA** (iSER) is supported for serving data for OpenStack block storage (Cinder). iSER-enabled Cinder storage requires an RDMA-capable NIC in both the compute hypervisors and the block storage servers exporting volumes of this type.

When using RDMA and SR-IOV-enabled NICs in an OpenStack private cloud, high levels of performance can be achieved from virtualized clients to a parallel file system. The multitenancy model of cloud infrastructure differs from the conventional multitenancy model of HPC parallel file systems, and this difference should be taken into account when connecting OpenStack compute instances with parallel file systems across the data center intranet, as we now explain. Conventional HPC platforms are multiuser environments, and user privileges and permissions are controlled through their user ID. In cloud-hosted infrastructure, it is standard practice to grant the tenant trivial access to root within their instances. When exporting file systems to cloud-hosted instances, provision should be made for potentially hostile clients with superuser privileges. Recent developments in Lustre have introduced Kerberos-based authentication for clients, which aims to resolve this issue. An alternative approach is to provision the creation of scratch parallel file systems for a tenant within their tenancy on the OpenStack cloud. In this way, a tenant is isolated from other cloud users and unable to subvert their access to a shared file system resource.

13.4 OpenStack Deployment

Even a default deployment of OpenStack is large and complex and is not normally deployed manually. The OpenStack project's rich software ecosystem includes a diverse range of automated systems for deployment and configuration. The OpenStack market has (broadly) converged on four approaches.

- *Turnkey systems.* Rack-scale appliances provide integrated private cloud compute and management.

- *Vendor-supported.* Linux distribution vendors are becoming dominant as OpenStack vendors. There are commercially-supported distributions of OpenStack from Canonical, Red Hat, and SUSE, and from OpenStack specialists such as Rackspace and Mirantis.

- *Community-packaged.* Freely-available Linux distributions such as CentOS and Ubuntu have community supported OpenStack packages.

- *Upstream code.* Some OpenStack deployments are assembled by using source code pulled directly from upstream source repositories, often deployed as containerized services.

Thus, users have considerable choice when selecting the means of deploying OpenStack. This choice provides for considerable flexibility when it comes to matching an organization's requirements, budget, and skill set to a suitable method of OpenStack deployment.

Deployment begins with several dedicated servers, networks, and disks. A conventional OpenStack deployment classifies these servers into distinct roles. On small-scale deployments, several roles may be combined to enable a scaled-down footprint, even down to a single node. These roles are listed below.

- *Compute hypervisors.* These are servers that run the client workloads in a virtualized environment. In addition to virtualized compute, services are usually required for implementing software-defined networking and storage.

- *OpenStack controllers.* Centralized OpenStack control services and data stores typically run on separate servers. These servers should be configured for supporting database IO patterns and a highly concurrent, high-throughput transactional workload.

- *Storage.* Many OpenStack deployments are underpinned by a Ceph storage cluster (although this is not a requirement). A wide range of vendors offer OpenStack connectivity for commercial storage products.

- *Networking.* Several approaches are available for implementing tenant networking in OpenStack, but the conventional and most established approach involves managing router gateways to tenant networks using software virtual switches on controller nodes. Since this can quickly become highly CPU-intensive (and a performance bottleneck for tenants), OpenStack deployments using Neutron IP routers often include dedicated servers for scaling up networking performance. See section 13.3 on page 285 for other strategies.

13.5   Example Deployment

We next describe a deployment of OpenStack carried out at the University of Cambridge research computing services. These incorporate many of the considerations outlined in section 13.3 on page 285 to provide a flexible but performant resource.

13.5.1   Hardware Components

As illustrated in figure 13.1 on the next page, our example deployment is on current-generation Dell Xeon-based servers equipped with current-generation Mellanox

Ethernet NICs that support remote direct memory access (RDMA) (using RoCE: RDMA over converged Ethernet) and SR-IOV. The servers are connected by a 50G Ethernet high-speed data network and a 100G multipath layer-2 Ethernet fabric, using multichassis link aggregation (MLAG) to achieve multipathing. The high-speed network uses Mellanox Ethernet switches. Separate 1G networks are used for power management and server provisioning and control. The system also uses a separate 10G network for the OpenStack control plane.

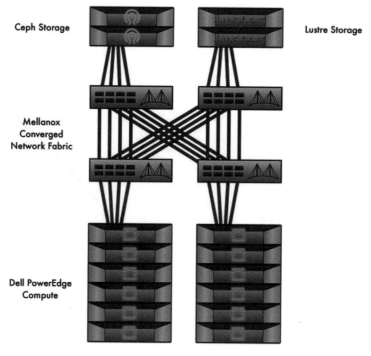

Figure 13.1: Hardware configuration used in our example OpenStack for science deployment.

Storage services are delivered by using a range of components. A high-speed storage service is implemented by using iSER and NVMe devices. A moderate-scale Ceph cluster is used to provide a tier of storage with greater capacity and resilience. Enterprise storage is provided by Nexenta. Outside of the OpenStack infrastructure, Intel Enterprise Edition Lustre is delivered to compute instances by using a data-center provider network.

13.5.2 OpenStack Components

We describe here a freely available, community-supported OpenStack deployment that uses the Community ENTerprise Operating System (CentOS) Linux distribution with OpenStack packages from the Red Hat Distribution for OpenStack (RDO). The servers are deployed with CentOS and OpenStack using TripleO, a tool for automated OpenStack deployment and management. The TripleO online documentation [49] provides a comprehensive guide to using the tool for OpenStack deployment. We focus here on adapting an OpenStack configuration to improve support for scientific computing workloads. To retain generality with other methods of deployment, we also describe the key components of the OpenStack configuration and how TripleO is used to realize that configuration.

13.5.3 Enabling Block Storage via RDMA

In order to use the iSER protocol, all associated storage servers and hypervisor clients must have both RDMA-capable NICs and the Open Fabrics (OFED) stack installed. The storage server manages the Cinder block storage volumes using LVM. The iSCSI protocol is configured as iSER in the OpenStack Cinder Volume driver configuration (`/etc/cinder/cinder.conf`) on the storage server, as follows.

```
[hpc_storage]
volume_driver=cinder.volume.drivers.lvm.LVMVolumeDriver
volumes_dir=/var/lib/cinder/volumes
iscsi_protocol=iser
iscsi_ip_address=10.4.99.3
volume_backend_name=hpc_storage
iscsi_helper=lioadm
```

To enable this using TripleO (which supports iSER configuration from OpenStack Ocata release), the following configuration is required.

```
parameter_defaults:
  CinderEnableIscsiBackend: true
  CinderIscsiProtocol: 'iser'
  CinderISCSIHelper: 'lioadm'
```

Once this configuration is deployed, iSER-enabled block storage volumes can be created and attached to compute instances, using the same interface as any other kind of Cinder volume.

13.5.4 Enabling SR-IOV Networking

Because of its circumvention of OpenStack security groups, SR-IOV is not suitable for use on externally accessible networks. SR-IOV requires hardware support in the NIC. This can be checked for in the product specs or with lspci -v. The PCI Vendor ID and Device ID of the NIC are needed for hypervisor configuration (the Mellanox ConnectX4-LX NICs used in this example configuration have IDs 0x15b3 and 0x1016, respectively). SR-IOV must be enabled in both BIOS and the Linux kernel. These additional kernel command-line boot parameters enable SR-IOV support in the Linux kernel as follows.

```
intel_iommu=on iommu=pt
```

SR-IOV networking requires configuration of both Nova and Neutron. Additionally, several SR-IOV virtual functions (VFs) must be created in advance, typically during system startup. On compute hypervisors, permission for PCI-Passthrough of the NIC VFs must be declared in Nova's configuration /etc/nova/nova.conf:

```
pci_passthrough_whitelist = [{"vendor_id": "15b3", \
                              "device_id": "1015", \
                              "physical_network": "hpc_network"}]
```

On OpenStack controller nodes, the Nova configuration file /etc/nova/nova.conf needs to be edited to configure the scheduler with an additional filter for scheduling instances according to availability of SR-IOV capable devices:

```
scheduler_default_filters = RetryFilter,AvailabilityZoneFilter,
    RamFilter, DiskFilter,ComputeFilter,ComputeCapabilitiesFilter,\
    ImagePropertiesFilter,ServerGroupAntiAffinityFilter, \
    ServerGroupAffinityFilter,PciPassthroughFilter
```

On OpenStack controller nodes, the SR-IOV network driver is configured in Neutron's configuration /etc/neutron/plugins/ml2/ml2_conf.ini with the PCI details of the NIC.

```
[ml2_sriov]
supported_pci_vendor_devs=15b3:1016
```

You must also edit /etc/neutron/plugins/ml2/ml2_conf.ini to set the VLAN range from which to allocate tenant networks, as follows:

```
[ml2_type_vlan]
network_vlan_ranges = hpc_network:1001:4000,external
```

```
parameter_defaults:
  NeutronBridgeMappings: "hpc_network:br50g,external:br10g"
  NeutronNetworkType: "vxlan,vlan"
  NeutronMechanismDrivers: "sdnmechdriver,openvswitch,sriovnicswitch"
  NeutronNetworkVLANRanges: "hpc_network:1001:4000,external"

  NovaComputeExtraConfig:
    neutron::agents::ml2::ovs::bridge_mappings: ['external:br10g']
    nova::compute::pci_passthrough:'"[{\"vendor_id\":\"15b3\", \
      \"device_id\":\"1015\",\"physical_network\":\"hpc_network\"}]"
    compute_classes:
      - ::neutron::agents::ml2::sriov

  controllerExtraConfig:
    nova::scheduler::filter::scheduler_default_filters:
      - RetryFilter
      - AvailabilityZoneFilter
      - RamFilter
      - DiskFilter
      - ComputeFilter
      - ComputeCapabilitiesFilter
      - ImagePropertiesFilter
      - ServerGroupAntiAffinityFilter
      - ServerGroupAffinityFilter
      - PciPassthroughFilter
    neutron::config::plugin_ml2_config:
      ml2_sriov/supported_pci_vendor_devs:
        value: '15b3:1016'
```

Figure 13.2: TripleO configuration for enabling SR-IOV support. As described in the text, PCI device details are configured, networking mechanisms are defined, and physical network connectivity is mapped. The Nova scheduler is extended with `PciPassthroughFilter`.

This TripleO configuration enables SR-IOV support for an internal network named `hpc_network`, using a defined range of VLANs. It also provides for Open vSwitch and VXLAN networking, for general-purpose and externally connected networking, as in figure 13.2.

Once OpenStack has been deployed with this configuration, SR-IOV network ports can be easily created either by using the OpenStack command-line interface or through an orchestrated deployment via OpenStack Heat. OpenStack refers to SR-IOV network ports as direct-bound ports.

When using the command line, four steps are needed in order to create a network for use with SR-IOV.

1. Create a VLAN network.

2. Assign an IP subnet to it.

3. Attach direct-bound (SR-IOV) ports to the network.

4. Create compute instances connected to those ports.

```
[neutron net-create hpc_net --provider:network_type vlan
neutron subnet-create --name hpc_net --gateway <gw>  \
      --dns-nameserver <dns>   --enable-dhcp hpc_net <cidr>
neutron port-create <net vlan uuid> --binding:vnic-type direct
nova boot --flavor <flavor> --image <image> \
      --nic port-id=<sriov port uuid> <name>
```

Enabling CPU pinning Associating virtual CPUs with physical cores can improve cache performance through improved locality. This kernel command-line boot parameter excludes a given range of CPUs from scheduling, effectively reserving those CPUs for virtualized workloads. On a 24-core system, we might assign four cores for hypervisor activities and reserve 20 cores for guest workloads.

```
isolcpus=4-23
```

Once the CPUs have been isolated from scheduling, they can be assigned in Nova for CPU pinning. In /etc/nova/nova.conf insert the following.

```
vcpu_pin_set = 4-23
```

The Nova configuration can also be specified in TripleO-driven deployments.

```
parameter_defaults:
  NovaComputeExtraConfig:
    nova::compute::vcpu_pin_set: 4-23
```

Enabling NUMA passthrough Further performance gains can be achieved by exposing the physical topology of processors, memory, and hardware devices. A guest operating system that is NUMA-aware can exploit this awareness to improve the efficiency of virtualized application workloads, in the same manner as for workloads running in a bare metal environment. Passthrough of NUMA topology is supported for current versions of KVM and libvirt, and (since OpenStack Juno) OpenStack. The release of KVM that ships with CentOS (7.3) requires updating to the version available in the CentOS-virt KVM repository. KVM 2.1.0 is the minimum required version for supporting NUMA passthrough.

The NUMA topology requested by a compute instance is defined by using additional properties of either a compute flavor or a software image. The OpenStack documentation on CPU topologies [36] describes how flavors and images may be configured to specify the underlying NUMA resources desired for supporting the workload. In a similar manner to SR-IOV support, NUMA awareness requires a filter in the Nova compute scheduler to ensure that available resources meet the stated requirements of the flavor of compute instance being scheduled. Scheduler filters are specified in /etc/nova/nova.conf by setting the scheduler_default_filters property, as in the following example.

```
scheduler_default_filters = RetryFilter,AvailabilityZoneFilter, \
            DiskFilter,ComputeFilter,ComputeCapabilitiesFilter, \
            ImagePropertiesFilter,ServerGroupAntiAffinityFilter, \
            ServerGroupAffinityFilter,PciPassthroughFilter, \
            RamFilter, NUMATopologyFilter
```

This TripleO config file deploys the Nova Compute Scheduler with NUMA awareness enabled.

```
parameter_defaults:
  controllerExtraConfig:
    nova::scheduler::filter::scheduler_default_filters:
        - RetryFilter
        - AvailabilityZoneFilter
        - RamFilter
        - DiskFilter
        - ComputeFilter
        - ComputeCapabilitiesFilter
        - ImagePropertiesFilter
        - ServerGroupAntiAffinityFilter
        - ServerGroupAffinityFilter
        - PciPassthroughFilter
        - NUMATopologyFilter
```

13.6   Summary

We have presented the key architectural components of the OpenStack system and illustrated the basic concepts required to deploy the OpenStack software. In keeping with the theme of this book, we have focused on scientific use cases and provided tips on how to optimize performance.

OpenStack provides the benefits of software-defined infrastructure, while minimizing associated performance overheads. Diverse approaches to OpenStack configuration and deployment yield a range of trade-offs between flexibility and performance. The rapid pace of OpenStack's evolution seems likely to ensure that as the platform matures, its value for scientific compute infrastructure management will become increasingly compelling.

13.7 Resources

Many OpenStack private clouds are deployed for research computing, and a significant proportion of research computing private clouds are deployed using freely available versions of OpenStack. The operators of these clouds depend on an open community for support.

OpenStack operators of all forms exchange information and experiences through the OpenStack Operators mailing list [39] and through regular meetups. The OpenStack Foundation established the Scientific Working Group [40] as a focal point for the specific interests of the research computing community. The Scientific Working Group is free to join and draws its membership from a global network of research institutions that are using OpenStack infrastructure to meet the requirements of modern research computing.

Chapter 14

Building Your Own SaaS

"Great services are not canceled by one act or by one single error."
—Benjamin Disraeli

We saw in part II how software can be made portable and reusable by encapsulating it within a virtual machine or container. Anyone with a cloud account can then run that code at a click of a mouse or button. Nevertheless, users of such software must be concerned with obtaining the latest software package, launching it on the cloud, verifying correct execution, and so forth. Each of these activities introduces cost and complexity.

Software as a service (SaaS) seeks to overcome these challenges by introducing a further degree of automation in software delivery. With SaaS, software is operated on the user's behalf by a SaaS provider. Users then access the software over the network, often from a web browser. They need not install, configure, nor update any software, and software providers only have to support a single software version, which they can update at any time to correct errors or add features. SaaS providers typically use techniques such as replication to achieve high reliability, so that the service is not liable to be "cancelled by one single error." SaaS advocates argue that this approach reduces both complexity for software consumers and costs for providers. Subscription-based payment schemes further reduce friction for consumers and enable providers to scale delivery with demand. As we saw in chapter 1, thousands of SaaS providers now operate software in this manner.

In this chapter, we examine how SaaS methods can be applied in science. We first dissect why it is that SaaS is conventionally defined as both a technology and business model. We then review approaches to on-demand software in science, including the popular concept of a science gateway, a form of remote software access that has primarily targeted research supercomputer systems for computation. We then look at more cloud native versions of the science gateway SaaS concept, using two examples to illustrate how you can build SaaS solutions for your own purposes. The first, **Globus Genomics**, provides on-demand access to bioinformatics pipelines. The second is the **Globus** research data management service, already introduced in chapter 11. While it so happens that both systems that we present here build on Amazon cloud services, similar implementation and deployment strategies can be followed on other public clouds. A key message is that leveraging cloud platform services can facilitate the creation of SaaS offerings that are reliable, secure, and scalable in terms of the number of users supported and amount of service delivered.

14.1   The Meaning of SaaS

Gartner defines SaaS as software that is "owned, delivered, and managed remotely by [a provider who] delivers software based on one set of common code and data definitions that is consumed in a one-to-many model by all contracted customers at any time, on a pay-for-use basis or as a subscription" [136]. Analyst David Terrar adds: "[and t]he application behind the service is properly web architected—not an existing application web enabled."

This definition is as much about business model as technology. From a technology perspective, it speaks to a delivery model: the provider runs a single version of the software, deployed so as to permit access over the Internet via web interfaces. By implication, the software is architected to scale to large numbers of consumers, and to enable multitenant operations, meaning that multiple consumers can access the software at the same time without interfering with each other. From a business model perspective, the definition speaks to software that consumers do not buy, as they would a home appliance, but instead pay per use, as they would a movie online, or subscribe to, as they would a newspaper.

This juxtaposition of technology and business model may seem odd, but in fact it is central to the success of SaaS, in industry at least. (The fact that some SaaS providers use advertising revenue rather than subscriptions to cover their costs does not change the essential economics.) In brief, the centralized operation model has allowed SaaS providers to slash per-user costs relative to software distributed via traditional channels, because there is, for example, no longer a need to support multiple computer architectures and versions. It has also greatly reduced barriers to access: most SaaS software is accessible to anyone with a Web browser, in contrast to much enterprise software that might require specialized hardware and expertise to install and run. These two factors mean that SaaS providers can deliver software to many more people at far lower costs than were previously possible: literally cents on the dollar.

While the cost of serving each new customer may be low, it is not zero. Furthermore, the upfront cost of establishing and operating a SaaS system involves significant fixed costs. (For example, 24x7 monitoring to ensure high availability.) The SaaS industry has determined that pay-per-use or subscription-based payment models are the best way to recoup these costs. Such approaches provide a low barrier to entry (anyone with a credit card can access a service) and mean that revenue scales linearly with usage. Thus, many users and per-user payments make SaaS sustainable by providing positive returns to scale: more users means more income that can pay for the scaled-up operations and/or reduce subscription charges, encouraging yet broader adoption.

14.2   SaaS Architecture

Leaving business model aside, we next introduce approaches that have proven successful for architecting and engineering SaaS systems.

In general, the goal of a SaaS architect is to create a system that can deliver powerful capabilities to many customers at low cost and with high reliability, security, and performance. This overarching goal allows for many tradeoffs: for example, between capabilities and reliability, or between optimizing for base cost or per-user cost. Nevertheless, some basic principles can be identified.

The most sophisticated SaaS systems are often architected using a microservice architecture, in which state is maintained in one or more persistent, replicated storage services, and computation is performed in short-lived, stateless services that can be rerun if necessary. This architecture provides for a high degree of fault tolerance and also facilitates scaling: more virtual machines can be allocated dynamically as load increases. The following example illustrates these principles.

Video rendering SaaS. A company called **Animoto** has had a lot of success with their video rendering service. You upload a set of photos; they create an animated video from theset images, with smooth transitions from picture to picture and musical accompaniment. They thus need to run many somewhat data- and computation-intensive tasks. As shown in figure 14.1 on the next page, their architecture uses Amazon S3 cloud object storage to hold all images and data; dynamically managed Amazon EC2 virtual machine instances to run web servers, data ingest servers,

---

rendering servers, and the like; and the Amazon SQS queuing service to coordinate between these activities, for example to maintain pending data ingest, analysis, and rendering tasks. At least that is how they described their architecture almost a decade ago; they likely use a richer set of services now. Animoto describes scaling to support 750,000 new users in three days by adding virtual machine instances [4].

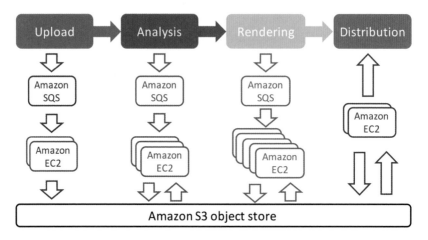

Figure 14.1: A schematic of the Animoto SaaS pipeline, showing the logical structure at the top and its realization in terms of Amazon services below.

Let us now dive more deeply into the process by which a SaaS system is created. Imagine that you developed your own video rendering application, myrender, that creates videos like those generated by Animoto, but from the command line, for example as follows.

```
myrender -i image-directory-name -o video-file-name
```

Your application runs nicely on your GPU-equipped workstation, but it is becoming too popular for you to handle the growing number of requests by hand. You share the source code, but people keep asking you to perform more and more tasks. You want to deliver this application to many people as SaaS. How might you proceed? Let us consider three alternatives.

In the first, you write a program that accepts user rendering requests via a web form, and for each request launches myrender on your workstation. However, this approach breaks down as the number of requests continues to grow: your workstation is overwhelmed. A second approach is to adapt your program to instead instantiate a cloud-hosted virtual machine instance for each request. The virtual machine image is configured to accept user images, run the `myrender` application, wait for the user to download the resulting video, and then terminate.

A third approach is to partition the `myrender` code into separate analysis, rendering, and assembly components; create a virtual machine or container for each component; and create or use a framework akin to that used by Animoto to dispatch appropriate tasks to the different virtual machines or containers, while scaling their numbers up and down in response to changing load. Each individual request is thus represented solely by a set of objects in object storage and transient requests in a queue, plus perhaps entries in other database tables.

Which approach is better? The first does not scale, so let us consider the second and third. The second is surely less work to implement than the third: you do not need to alter your application code at all. However, the fact that each user rendering request is run on a separate virtual machine instance also has disadvantages. In particular: (1) Cost: You pay for each virtual machine, even when it is not fully occupied, for example because it is waiting on user input. (2) Speed: Each user request is processed sequentially, one image at a time. Opportunities for parallelism, for example to process all images in parallel, cannot be exploited without changes to the application. (3) Reliability: Failure of a virtual machine instance at any time results in the loss of all work performed by that instance to that time. Recovery requires restarting from the beginning. The third approach, on the other hand, requires more work up front but can provide big improvements in cloud cost, execution speed, and service reliability.

The third approach is commonly referred to as a **multitenant architecture** because all requests are fulfilled by the same (albeit elastically scalable) set of cloud resources, each of which may host different user requests at different times, and indeed multiple requests at the same time. A vital issue in multitenant systems is ensuring **isolation** among different users. It is the approach that is adopted in the vast majority of large SaaS systems.

14.3   SaaS and Science

The scientific community has a long history of providing online access to software. After all, the original motivation for the ARPANET, precursor to today's Internet, was to enable remote access to scarce computers [257]. With the advent of high-speed networks followed by the World Wide Web, many such experiments were conducted [126, 239]. One early system, the Network Enabled Optimization Server (NEOS) `neos-server.org`, has been in operation for more than 20 years [105], solving optimization problems delivered via email or the web.

The term **science gateways** has become increasingly often used to denote a system that provides online access to scientific software [261]. In general, a science gateway is a (typically web) portal that allows users to configure and invoke scientific applications, often on supercomputers, providing a convenient *gateway* to otherwise hard-to-access computers and software.

The impact of such systems on science has been considerable. For example, the MG-RAST metagenomics analysis service `metagenomics.anl.gov`, which provides online access to services for the analysis of genetic material in environmental samples [199], has more than 22,000 registered users as of 2017, who have collectively uploaded for analysis some 280,000 metagenomes containing more than 10^{14} base pairs. That is a tremendous amount of science being supported by a single service! Other successful systems, such as CIPRES [201], which provides access to phylogenetic reconstruction software; CyberGIS [185], for collaborative geospatial problem solving; and nanoHUB [172], which provides access to hundreds of computational simulation codes in nanotechnology, also have thousands of users and correspondingly large impacts on both science and education. A recent survey [176] provides further insights into how and where science gateways are used.

While it is hard to generalize across such a broad spectrum of activities, we can state that the typical science software service has some but not all of the properties of SaaS as commonly understood. First, from a technology perspective: Most such services commonly make a single version of a science application available to many people, and many leverage modern web interface technologies to provide intuitive interactive interfaces. Some also provide REST APIs and even SDKs to permit programmatic access. On the other hand, many are less than fully elastic, due to a need to run on specialized and typically overloaded supercomputers, and few are architected to leverage the power of modern cloud platforms. Thus, they handle modest numbers of users well, but may not scale.

From a business model perspective, few science software systems implement pay-by-use or subscription-based payment schemes. Instead, they typically rely on research grant support and/or allocations of compute and storage resources on scientific computing centers. This lack of a business model can be a subject of concern, because it raises a question about their long-term sustainability (what happens when grants end?) and also hinders scaling (an allocation of supercomputer time may be enough to support 10 concurrent users, but what happens when demand increases to 1000 concurrent users? 10,000?).

We next use two example systems that have each taken a different approach to science SaaS from both technology and business model perspectives: Globus Genomics and the Globus service.

14.4   The Globus Genomics Bioinformatics System

Globus Genomics [187] `globus.org/genomics`, developed by Ravi Madduri, Paul Davé, Alex Rodriguez, Dina Sulakhe, and others, is a cloud-hosted software service for the rapid analysis of biomedical, and in particular next generation sequencing (NGS), data. The basic idea is as follows: a customer (individual researcher, laboratory, community) signs up for the service. The Globus Genomics team then establishes a service instance, configured with applications and pipelines specific to the new customer's disciplinary requirements. Access to this instance is managed via a Globus Group. Any authorized user can then sign on to the instance, use its Galaxy interface to select an existing application or pipeline (or create a new pipeline), specify the data to be processed, and launch a computation that processes the specified data with the specified pipeline. Computational results can be maintained within the instance or, alternatively, returned to the user's laboratory for further processing or long-term storage.

Figure 14.2 shows Globus Genomics in action. We see its Galaxy interface being used to display a pipeline commonly employed for the analysis of data from an RNA-seq experiment, which is a method for determining the type and quantity of RNA in a biological sample [99]. By providing research teams with a personal cloud-powered data storage and analysis "virtual computer," Globus Genomics allows researchers to perform fully automated analysis of large genetic sequence datasets from a web browser, without any need for software installation or indeed any expertise in cloud or parallel computing. In one common use case, a researcher sends a biological sample to a commercial sequencing provider, has the resulting data communicated over the network to cloud storage (e.g., Amazon S3); and then accesses and analyzes the data by an analysis pipeline running within Globus Genomics.

14.4.1   Globus Genomics Architecture and Implementation

As shown in figure 14.3 on page 307, the Globus Genomics implementation comprises six components, all deployed on a single Amazon EC2 node: Galaxy and web server for workflow management and user interface; HTCondor and an elastic provisioner for computation management; and Globus Connect Server (GCS) and shared file system for data management. These services themselves engage other cloud services, notably Globus identity and data management services (see section 3.6 on page 51) for user authentication and to initiate data transfers; Amazon EC2 to create and delete the virtual machine instances on which user computations run; and the Amazon Relational Database Service and Elastic File System for

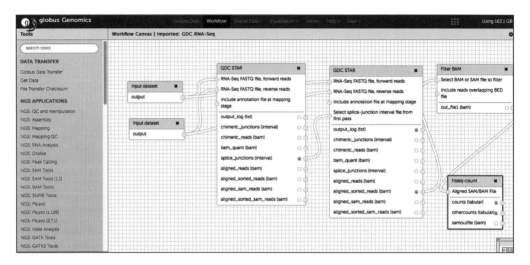

Figure 14.2: The Galaxy web interface used by Globus Genomics, showing an RNA-seq pipeline that allows a researcher to detect various features in experimental samples.

storing user data that needs to persist over time. We describe each element in turn.

The **Galaxy** system [141] supports construction and execution of workflows. A user signs on to the cloud-hosted Galaxy, using campus credentials thanks to integration with Globus Auth [247]. They can then select an existing workflow or create a new one, identify data to be processed, and launch computational tasks. The **web server**, an integral part of the Galaxy system, serves the Galaxy user interface to Globus Genomics users. Users need only a web browser to access Globus Genomics capabilities.

The **HTCondor** system [243] (see section 7.7 on page 128) maintains a queue of tasks to be executed, dispatches tasks from that queue to available EC2 worker nodes, and monitors those tasks for successful completion or failure. The **elastic provisioner** manages the pool of worker nodes, allocating nodes of the right type for the tasks that are to be executed, increasing the number of nodes when the HTCondor queue becomes long, and de-allocating nodes when there is little or no work to do. The elastic provisioner is designed to use spot instances (see section 5.2.2 on page 77) where possible, in order to reduce costs.

Globus Connect Server (GCS), as discussed in section 3.6 on page 51, implements the protocols that the Globus cloud service uses to manage data transfers between pairs of Globus Connect instances. (The related Globus Connect Personal service is designed for use on single-user personal computers.) We can think of this component as being equivalent to the agent that runs on your personal computer to interact with the Dropbox file sharing system, although GCS supports specialized high-speed protocols.

> **AWS Batch** `aws.amazon.com/batch`. Amazon recently released this job scheduling and management service, which they indicate can run hundreds of thousands of batch computing jobs on the Amazon cloud, dynamically provisioning compute resources of types (e.g., CPU or memory optimized instances; EC2 and Spot instances) and numbers required to meet the resource needs of the jobs submitted. If AWS Batch behaves as advertised, it might obviate the need for the HTCondor and elastic provisioner components of the Globus Genomics solution.

Finally, the **shared file system** uses the Network File System (NFS) to provide a uniform file system name space across the manager and worker nodes. This mechanism simplifies the execution of Galaxy workflows, which are designed to run in a shared file system environment.

Globus Genomics uses Chef `chef.io` for Configuration Management, allowing service components to be updated and replaced without any error-prone manual configuration steps [186]. It uses Chef recipes to encode the following steps.

1. Provision an Identity and Access Management (IAM: see section 15.2 on page 319) user under the customer's Amazon account, with a security policy that allows the IAM user to create and remove AWS resources and perform other actions required to set up and run a production-grade science SaaS.

2. Provision an EC2 instance with HTCondor software, configured to serve as the head node for an HTCondor computational cluster; Network File Server software for data sharing with computational cluster nodes; a NGINX web proxy and WSGI Python web server; a Galaxy process; a Globus Connect server for external access to the network file system, configured for optimal performance; Unix accounts for the administrators; security updates/patches; and Domain Name System (DNS) support, via Amazon's **Route 53** service.

3. Provision the following additional components:

   (a) An Amazon **Virtual Private Cloud** with appropriate network routes between the head nodes and compute nodes;

   (b) An EBS/EFS-based network file system with optimized I/O configuration;

(c) An elastic provisioner along with network configurations to support the creation of spot instances across multiple Availability Zones;

(d) A read-only network volume configured with the tools and pipelines required for a specific scientific domain, and with reference datasets that may be used by analysis pipelines;

(e) Monitoring of the health of various system components, generating alerts as required;

(f) An Amazon **Relational Database Service** database for persisting the state of application workflows; and

(g) Identity integration with Globus, and groups configured for authorization of user accesses to the instance.

14.4.2 Globus Genomics as SaaS

Globus Genomics has many SaaS attributes. From a technology perspective, it is accessible remotely over the network, runs a single copy of its software (Galaxy, the various genomics analysis tools and pipelines), and leverages cloud platform services provided by Amazon and Globus for scalability and to simplify its implementation.

Globus Genomics is not multitenant: it creates a separate instance of the system (the manager node in figure 14.3) for each customer, rather than having one scalable instance serving all customers. This characteristic of the Globus Genomics system is not a problem for users: indeed, a single-tenant architecture may appear preferable to some due to (at least an appearance of) increased security and the clean, transparent billing for Amazon cloud charges that it allows. However, single tenancy increases costs for the Globus Genomics team over time, as each new customer requires the instantiation of a complete new Globus Genomics configuration, increasing Amazon usage and other operations costs, and different customers cannot share compute nodes.

We note also that the implementation does not guard against failure of the node on which the manager logic runs. The Globus Genomics team can detect such failures and restart the service, but the failure is not transparent to users. One solution would be to re-engineer the system to leverage more microservices, as we described for Animoto above.

From a business model perspective, Globus Genomics also has SaaS characteristics in that its use is supported by a subscription model. A lab or individual user signs up for a Globus Genomics subscription that covers the base cost of operating their private instance. As part of the configuration of a Globus Genomics instance,

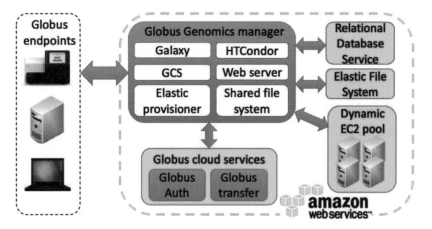

Figure 14.3: The Globus Genomics system dispatches tasks to a dynamically instantiated HTCondor pool, with virtual nodes added and removed by the elastic provisioner in response to changing load.

Amazon account details are provided so that resources consumed by any users granted access to that instance can be charged to that account.

14.5   The Globus Research Data Management Service

Globus research data management service!as software as a service|(

The two limitations noted in our discussion of Globus Genomics illustrate tradeoffs that frequently arise when developing cloud-hosted SaaS, particularly in science. Multitenancy and microservice architectures tend to reduce costs and increase reliability, but can increase up-front costs. In our second example, we describe a system that is more cloud native, namely the Globus research data management service introduced in section 1.5.4 on page 15.

Globus, developed at the University of Chicago since 2010, leverages software-as-a-service methods to deliver data management capabilities to the research community. As shown in figure 14.4 on the next page, those capabilities, which include data transfer, sharing, publication, and discovery as well as identity and credential management, are implemented by software running on the Amazon cloud. Globus Connect software deployed on file systems at research institutions and on personal computers enable those systems to participate in the Globus file sharing network. REST APIs support programmatic access, as we have described in chapter 11 and in the preceding Globus Genomics section.

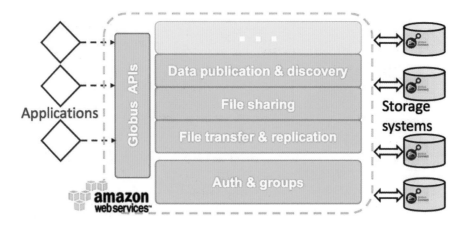

Figure 14.4: Globus SaaS provides authentication and data transfer, sharing, publication, and discovery capabilities, accessible via APIs (left) and web clients (not shown). Globus Connect software on storage systems enables access to data in many locations.

Globus is popular because researchers and developers of research tools alike can hand off to the Globus service responsibility for otherwise time-consuming tasks, such as babysitting file transfers. For example, consider a researcher who wants to transfer data from site A to B. With Globus, the researcher can simply make a request to the cloud-hosted service, via API or web interface. The Globus service then handles user authentication, negotiation of access at sites A and B, configuration of the transfer, and monitoring and control of the transfer activity.

Because many important projects depend on Globus for authentication, authorization, data access, and other purposes, high availability is essential. Thus the Globus implementation leverages public cloud services to replicate state data in multiple locations, operate redundant servers with dynamic failover, monitor service status, and so forth. Table 14.1 provides a partial list of the Amazon services used by Globus.

14.5.1 Globus Service Architecture

The Globus SaaS is broken down into logical units of services. Each service comprises three key components: a **REST API**, a set of one or more **backend task workers**, and a **persistence layer**. Additional components may be needed for some services, and some components may be colocated to save cost or complexity. Having this common breakdown, and exposing the services to one another only via their REST APIs, provides several key properties that allow different parts of

Table 14.1: Some of the Amazon cloud services used in the Globus SaaS implementation.

Service	Use made by Globus
EC2	Provide high availability instances of Globus services; serve web APIs; run background tasks; internal infrastructure
RDS	Store Globus service state with high availability and durability
DynamoDB	Store Globus service state with high availability and durability
VPC	Establish private Amazon cloud with secure virtual network
ELB	Direct client requests to an available service instance
S3	Store state of in-progress tasks, service data backups, static web content
IAM	Manage access to Amazon resources within Globus
CloudWatch	Monitor the status of Globus resources
SNS	Simple Notification Service to send notifications to Globus staff
SES	Simple Email Service to deliver email to users

the SaaS to scale independently of one another.

Globus REST APIs are typically deployed on EC2 instances. All logic used to handle REST API requests is performed synchronously: any asynchronous or long-term activity requested of a service is handled by creating records of the desired activity in the persistence layer. The REST API handlers do not wait for these actions to be completed, but simply register the desired activity in persistent storage and terminate. Further processing is then handled by the backend task workers, which either poll the persistence layer or are notified by API workers. This approach gives the REST API instances the powerful property of being **stateless**: their contents on disk and in memory are ephemeral, and any REST API instance can process any request. As a result, these microservices can scale up and down more or less trivially, allowing the Globus team to add or remove capacity to serve APIs in direct proportion to observed system load. So long as the backend task workers rely on the persistence layer and are also stateless, they can scale up or down just as easily.

Globus employs the same public-facing REST APIs for all internal communications between services and thus no two Globus services are tightly coupled. Each service can scale, rearrange infrastructure, and alter core service components without impacting one another at all. This separation of concerns is key to Globus operations, and allows improvements to be made safely, easily, and frequently.

The persistence layer is implemented on Amazon storage services, leveraging their replication across availability zones for fault tolerance and creating periodic remote snapshots for disaster recovery. Globus uses, in particular, S3 and the PostgreSQL Relational Database Service (RDS). One service uses DynamoDB. The various system components are encapsulated in Virtual Private Clouds (VPCs), which allow for the provisioning of logically isolated sections of the Amazon cloud within which Amazon resources can be launched in a managed virtual network.

14.5.2 Globus Service Operations

In addition to segmenting the SaaS into component services, Globus leverages a number of common practices across all of these services to maintain them uniformly. By so doing, operational and infrastructural improvements are applied across the entire product offering and their effects are amplified. Internal components that are thus shared across multiple or all Globus services include service health and performance monitoring; continuous integration and continuous delivery pipelines; security monitoring and intrusion detection; log aggregation; configuration management; and backups of disk, database, S3, and other storage.

Globus research data management service!as software as a service|)

14.6 Summary

Our goal in providing this brief review of SaaS methods is to give you a framework for your thinking on software as a service and its role in science. True SaaS, realized as cloud-hosted software with support from pay-for-use or subscriptions, can address in a convenient manner the three major challenges of science software, namely usability, scalability, and sustainability. But a certain scale of use is required to justify the costs of multitenant architecture, and not all software will generate the interest and subscription income required to support that scale of use. The science community will surely learn a lot more about the pros and cons of SaaS in the next few years.

14.7 Resources

Dubey and Wagle provide a somewhat dated but still excellent overview of software as a service [113]. Fox, Patterson, and Joseph's *Engineering Software as a Service* [128], designed to accompany their EdX online course, provides in-depth discussion of many issues that arise when building software as a service.

Part V

Security and Other Topics

Building your own cloud
What you need to know
Using Eucalyptus
Part IV   Using OpenStack

Security and other topics
Securing services and data
Solutions
History, critiques, futures   **Part V**

Part III   **The cloud as platform**
Data analytics   Streaming data   Machine learning   Research data portals
Spark & Hadoop   Kafka, Spark, Beam   Scikit-Learn, CNTK,   DMZs and DTNs, Globus
Public cloud Tools   Kinesis, Azure Events   Tensorflow, AWS ML   Science gateways

Managing data in the cloud
File systems
Object stores
Databases (SQL)
NoSQL and graphs
Warehouses
Part I   Globus file services

Computing in the cloud
Virtual machines
Containers – Docker
MapReduce – Yarn and Spark
HPC clusters in the cloud
Mesos, Swarm, Kubernetes
Part II   HTCondor

Part V:
Security and Other Topics

We have described in the 14 preceding chapters how to use cloud services to store data and perform computations. We have shown how to use platform services for data analytics, streaming data and machine learning, and how you can build your own services using the Globus platform. In the two remaining chapters, we discuss one essential topic, namely security, and discuss historical perspectives, contemporary critiques of cloud, and futures.

Throughout the preceding chapters we touched only lightly on security. Now that you have a more complete view of cloud features and capabilities, it is time to return to the subject of security in a more comprehensive manner. In chapter 15, we first address the issue of security responsibility: which security and privacy issues are the job of the cloud provider and which belong to you. One way to understand this bifurcation is that the cloud provider is responsible for the security *of* the cloud, while you are responsible for what you do *in* the cloud.

We then address three central security topics: securing data that you move to the cloud, securing access to the virtual machines and containers that you create in the cloud, and using cloud services in a secure manner. We discuss user authentication and authorization: determining who someone is and what they are allowed to do. We described Globus authentication and authorization mechanisms in chapter 11; here, we discuss additional approaches, including Amazon's and Azure's role-based access control mechanisms. We also cover virtual machine and container security and how to secure the cloud software services that you create.

In chapter 16, we explore the history that led to the current cloud environment. We also return to the pros and cons of public cloud computing, a topic that we first addressed in section 4.4 on page 68 but now revisit with a broader perspective. We conclude with a look at future trends in cloud data center architecture and newly emerging approaches to cloud software.

Chapter 15

Security and Privacy

"The ultimate security is your understanding of reality."

—H. Stanley Judd

We leave security to the end of the book not because it is an afterthought but because it is a cross-cutting concern. Indeed, it is important for every aspect of how you work with the cloud. In this chapter we start with the general question of what the cloud provider provides you as a security baseline and what aspects of security are your responsibility. Then we turn to the question of how best to protect your data, computations, and services. We showed in part I of this book that managing data in the cloud is relatively easy, being supported by both intuitive cloud portals and programming APIs. Here we look deeper into the best practices of data protection. Similarly, we talked in part II about computing in the cloud using VMs, containers, and clusters of both. However, we said virtually nothing about securing those VMs or managing the networks that connect them. In this chapter we introduce you to critical issues that you need to consider when deploying these computing resources. Finally, we take a brief look at how to use higher-level services securely.

15.1   Thinking about Security in the Cloud

One frequently cited reason for not using the cloud is concern about security. This concern is understandable: after all, data on your own computer are clearly under your control, while data on the cloud are somewhere unknown. But does this difference in location mean that your data are more or less secure? In general, your data should be *more* secure in the cloud than on your personal computer. This observation may appear paradoxical, but remember: operators of cloud services are information technology professionals whose livelihoods depend on preventing intrusions. The same cannot be said of most, if not all, readers of this book.

But just because cloud data centers are secure does not necessarily mean that your data are secure. In fact, your concerns about cloud security may be justified if you do not take proper care. So let's look at some best practices for cloud security.

The first point to remember is that any time that your computing infrastructure is exposed to the Internet, security is a concern. That is, any service that accepts and processes messages communicated over the network is potentially vulnerable to attack. This caution applies to your personal, home, laboratory, and institutional research systems as well as the cloud. Have you installed all the latest security patches on your personal machines? Is your institution's data center well managed? If you have secured your own systems, you have closed off the most common way for an intruder to access your cloud resources: the biggest risk in cloud computing is probably someone breaking into your personal computer and then accessing your cloud resources from there.

Securing your personal computers is out of scope for this book, so we now turn to cloud security issues proper. Three main areas merit your attention, each of which we consider in a subsequent section.

1. Secure data that you move to the cloud

2. Secure access to the virtual machines and containers that you create

3. Use cloud software services in a secure manner

In each of these three areas, much of the responsibility for the security of your cloud data rests with you. As shown in figure 15.1 on the following page, the cloud provider manages security *of* the cloud, implementing and operating security measures that protect the cloud infrastructure. However, security *in* the cloud is your responsibility: you are the one who defines the security mechanisms that you deploy to protect your own content, platform, applications, systems, and networks. The situation is much the same as if your applications were running in an on-site data center, except that cloud data centers raise unique concerns.

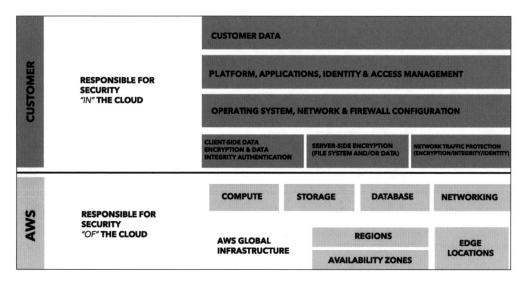

Figure 15.1: Shared responsibility model, as defined by Amazon [55].

Bad security can be expensive: A story of unexpected charges. This cautionary tale, unfortunately not unprecedented in cloud computing, is from Quora:

> **My AWS account was hacked and I have a $50,000 bill. how can I reduce the amount I need to pay?**
> For years, my bill was never above $350/month on my single AWS instance. Then over the weekend someone got hold of my private key and launched hundreds of instances and racked up a $50,000 bill before I found out about it on Tuesday.

What happened? The cause here, and it turns out a surprisingly common cause for such events, is that the user had included an Amazon access key (see section 3.2 on page 38) in code that the user pushed to a public GitHub repository. The bad actor found the key via a scan of publicly accessible GitHub repositories and then used it to perform BitCoin mining.

Such stories usually have a happy ending, in that Amazon generally seems willing to waive such charges if they are shown to result from fraud. But clearly you do not want this misfortune to happen to you. The most important step that you can take to avoid it is not to expose your Amazon access key(s), whether on GitHub or elsewhere. Particularly if you are using Amazon only for simple testing, you may not think of an access key as being especially valuable. But as this example shows, it is. You need to protect it as if it were worth many thousands of dollars.

Many proactive steps can help you guard against unexpected charges. Amazon describes a long list of best practices that you are well advised to follow [5]. Important ideas are: protect your keys, for example by enabling multifactor authentication; never share keys with other users (instead, create new IAM users to which you grant required permissions); follow the principle of least privilege when creating IAM users (i.e., configure them to be able to perform only the actions that you expect them to perform, such as only read storage or only access storage in certain regions); and monitor usage and billing.

The Amazon **CloudWatch** service can be used to monitor a variety of metrics, including billing, and to define thresholds upon which email alerts are generated. Make sure to read your email! Another useful service, not for detecting illicit behavior but for recovering afterwards, is Amazon **CloudTrail**. This service allows you to obtain a history of Amazon API calls and related events for your account.

On Azure, you can use the **Azure security center** service to obtain an analysis of all data and compute resources that you have deployed on your account. This service can, for example, scan each of your data containers to determine their encryption and access status. The Azure **Threat Analytics** service is designed to detect abnormal behavior, malicious attacks, and other security issues in your environment. Azure also provides tools to enable **application whitelisting**, by which you declare which applications are allowed to access your resources.

Table 15.1 provides a complementary perspective to that depicted in figure 15.1, showing the level of shared responsibility for security across different levels of services and for software as a service, platform as a service, infrastructure as a service, and on-premises deployments. It shows for each level of service which responsibilities are the cloud provider's, which are yours, and which are shared.

In an on-premises deployment, all responsibility lives with you—or, hopefully, your system administrators. On the cloud, responsibilities shift increasingly from you, or whoever is managing your cloud services for you, to the cloud provider,

Table 15.1: Microsoft's shared responsibility model [235].

	SaaS	PaaS	IaaS	On prem
Data classification & accountability	You	You	You	You
Client & endpoint protection	Shared	You	You	You
Identity & access management	Shared	Shared	You	You
Application-level controls	Provider	Shared	You	You
Network controls	Provider	Provider	Shared	You
Host infrastructure	Provider	Provider	Shared	You
Physical security	Provider	Provider	Provider	You

as you move up the stack from IaaS to SaaS. At the highest level, you are always responsible for ensuring that your data are correctly identified, labeled, and classified. Client and endpoint protection, that is, ensuring that the devices that connect to cloud services are correctly configured, are also usually tasks for which you are responsible. Identity and access management are solely your responsibility when you are working with IaaS, as discussed earlier; for SaaS and PaaS, the provider often handles access control, subject to policies that you define.

Responsibility for application-level controls, such as applying up-to-date OS patches, configuring application software correctly, and ensuring that application software has no security holes, rests solely with the SaaS provider, who has full control over their application software. In the PaaS case, it is a shared responsibility of you and the provider, since application code is a mix of your code and PaaS code. In the IaaS case, responsibility rests solely with you.

The bottom three levels in the table are concerned with correctly configuring networks and providing necessary controls, such as VPNs; correctly configuring virtual machines, containers, storage systems, and other infrastructure elements; and (a topic that you may never have thought of, but one that is certainly important for sensitive data) addressing the physical security of the devices on which applications run. Responsibilities for these elements rest solidly with the provider in the case of SaaS and PaaS; for IaaS, responsibilities are shared.

15.2 Role-based Access Control

In academic settings, faculty commonly want to allow trusted postdocs and students to access their public cloud account, but in a way that grants those other individuals only restricted rights. A security construct called a **role** is widely used in public clouds for such purposes. Each time an account owner adds another individual to a cloud account, the owner specifies the role(s) that the new user has; each role defines something that the user is authorized to do.

In the case of Azure, the **role-based access control** (RBAC) [146] system allows you to control how different parties use resources under your account. Each new user must have a role: either a general role like "Contributor" or a more specific role like "Data Lake Analytics Developer" or "SQL DB Contributor." For example, a Contributor can use the resources but cannot grant access to another user; a Data Lake Developer can use only data lake services. The account manager can also monitor the usage made of your resources by different authorized individuals.

In the case of the Amazon cloud, the **Identity and Access Management** (IAM) service [52] that we introduced in section 7.6 on page 110 and discuss further below, provides similar capabilities. You can create a variety of different IAM roles and assign those roles to users, applications, and services. An IAM role defines who the user is and what that user is authorized to do. It also gives you a way to monitor the use by holders of different roles. Google's cloud uses a similar IAM system [1], as well as a system of access control lists.

15.2.1 Sharing Secrets among Containers in a Cluster

As we discussed in section 7.6.2 on page 112, containers present additional security issues, especially when the container instance is a stateless microservice that interacts with other services, because many instances of this container may then be stopped and started as needed. The problem one encounters is that these container instances may need to access various secrets, such as the API keys, identities, and passwords of the services that they invoke. While you can pass the keys to individual instances from the command line, that approach does not work for instances that are managed dynamically. Leaving the keys in the container Dockerfile is not secure, because they are then embedded in the Docker image.

As we have shown in section 7.6.4 on page 114, Amazon's IAM role system solves this problem for the Amazon container service, ECS. RBAC solves the similar problem for Azure. When using the Docker Swarm services to manage a collection of containers, you can use the `docker secret create` command to send Docker a secret that is sent securely to the swarm manager, where it is encrypted. When an authorized microservice is launched, the Swarm manager sends that secret to the microservice, where it is stored in an in-memory file system that is deleted when the microservice container is deleted [181].

15.3 Secure Data in the Cloud

Commercial cloud vendors operate highly secure data centers. While intrusions are certainly possible, such failures in operational security appear to be extremely rare. Thus, the two principal vulnerabilities for data in the cloud are, first, data in transit to and from the cloud and, second, unauthorized access due to failure by the user to set proper access permissions. (We do not discuss the case of data that are given up by a cloud vendor because of government court orders. Here we encounter complex international legal issues and national data sovereignty laws that are well beyond the scope of this book and the expertise of its authors. As far as we know, entities such as the National Security Agency (NSA) do not have backdoor access to cloud data centers. But what do we really know? If your level of paranoia about the NSA's interest in your data is not too high, read on.)

15.3.1 Secure Data in Transit

A potential weak link is in the Internet between you and the data center. For example, when you create a VM, you can specify which ports you want to leave open; if you are careless, you may leave some of these open to attack. The Python SDKs that we introduced in chapter 3 use **Transport Layer Security**: previously Secure Socket Layer, or SSL) connections (or HTTPS) to transfer your data. This is the same security mechanism that you use when you interact with online banking services. Experts consider it to be secure, although as we see below, they also recommend encrypting sensitive data for additional protection.

If moving data with Globus Transfer, you can request that data be encrypted prior to transfer. When using the Python SDK, you need only to set `encrypt_data=true`. In addition, Globus endpoints can be configured to force encryption for all transfers involving that endpoint, whether as source or destination. This option is always enabled for Amazon S3 endpoints, for example, as indicated in the fine print at the bottom of figure 3.8 on page 53.

15.3.2 Control Who Can Access Your Data

A second potential source of unwanted access is incorrectly configured access controls. When you upload data to your storage accounts, you are responsible for managing who and what can access those data. As described previously, role-based access control can allow you to restrict access from your team or your services to the data storage system. However, you must use different mechanisms to restrict access from external collaborators or the public to specific buckets. Typically, you are able to control whether data are accessible to all, to no one except yourself, or to only named individuals. A misconfigured access control specification (or, as discussed previously, improper release of a key) can easily result in the wrong people seeing your data.

Fortunately, access controls are easy to configure. In the case of Azure blob and table storage, the storage account has two associated keys, prosaically named `key1` and `key2`. We typically think of `key1` as the **master key**; that is what we used in the APIs, as discussed in section 3.3 on page 42. You should never share the master key with others. As we have previously described, you can give `key2` to collaborators, who then have full access to the storage account. You can regenerate either key if you want to terminate access. Within a storage account you create containers, and for each container you can grant different types of public access: no public access, public read access to all blobs in the container including listing them, or public access to blobs by name only. If you want to provide just a single individual with access to a container, you can use a shared access signature (SAS). This is a powerful mechanism for granting limited access to objects in your storage account to others, without having to expose your account key. Generating a SAS signature or setting these access controls is easy to do from the portal or the Azure SDK or from the Azure Storage Explorer running on your PC or Mac.

Amazon and Google have similar capabilities, differing only in the details, for managing access to their cloud storage

The Globus Auth service and API described in chapter 11 provide powerful authorization mechanisms that are used by Globus Sharing, for example, to allow user control over access to data at Globus endpoints.

15.3.3 Encrypt Your Data

You may sometimes want to go beyond the security provided by access controls. In particular, if disclosure of your data would be especially damaging, as when dealing with sensitive data pertaining to human subjects (see section 15.3.4 on the following page), you may wish (or be required) to ensure that those data are **encrypted at rest**. In other words, you want to ensure that your data are always encrypted when on cloud storage and are decrypted only when they need to be read, retrieved, or used for computation. In so doing, you can protect against disclosure due to mistakes made when setting access controls or breaches in cloud data-center security. Amazon and Azure both support two ways to encrypt data at rest: server-side encryption and client-side encryption.

Server-side encryption allows you to ask that the cloud vendor automatically encrypt data on arrival in the cloud and then decrypt that data automatically each time that you access them. For example, Amazon S3 allows you to request, when uploading data to S3, that server-side encryption be performed; Amazon then performs that encryption (and decryption on subsequent accesses) transparently. In Python, you simply add a third line to the code on page 41, as follows.

```
# Upload the file 'test.jpg' into the newly created bucket
s3.Object('datacont', 'test.jpg').put(
    Body=open('/home/mydata/test.jpg', 'rb'),
    ServerSideEncryption='AES256')
```

Amazon manages keys for you, encrypting each object with a unique key and encrypting that key itself with a master key that it regularly rotates. (A variant permits you to provide your own key with each upload and access request, so that the key is on Amazon computers only while being used for encryption or decryption.) As a further safeguard, you can obtain access to an audit trail of when your key was used and by whom. So as long as you trust Amazon to manage and apply your keys appropriately, this approach is highly secure.

Azure **Storage Service Encryption** provides similar capabilities for the Azure Blob service; the Google Cloud Datastore service has similar functionality. The services differ somewhat in how they allow users to control the application of server-side encryption. Amazon allows the user to require that all data uploaded to a container be encrypted; however, the encryption request must still be made on individual uploads, as indicated previously. (An attempt to upload data without the encryption parameter then raises an error.) Azure allows the user to enable encryption at the level of a storage account, a construct that we introduced in section 3.3 on page 42; once enabled, all data uploaded to that account are encrypted. Google Cloud Datastore always encrypts.

Client-side encryption is useful when you want to ensure that the cloud provider never has access to your unencrypted data. Amazon and Azure both provide tools that you can use to encrypt data before they are sent over the wire. You might use these tools, for example, to create a secure backup of data otherwise maintained in on-premises storage, particularly if regulatory requirements prevent unencrypted data from leaving your premises. But note that you are then responsible for preserving the keys (as you are with server-side encryption, if you provide the keys): if you lose a key, the data that it encrypted are also lost.

15.3.4 Complexities of Sensitive Data

If your work involves access to personal health data or other sensitive information, then you are likely subject to various rules and regulations that will affect whether and how you can use cloud resources. For example, in the U.S., work with **personal health information** (PHI) must comply with the provisions of the **Health Insurance Portability and Accountability Act** (HIPAA) and in particular its **Security Rule**, which mandates administrative, physical, and technical safeguards for electronic PHI. The processes by which a particular institution and application are deemed to be HIPAA compliant are complex and beyond the scope of this book. The important takeaway points are that (1) the major commercial cloud vendors can all satisfy HIPAA physical security standards, but (2) this does not mean that you can just put HIPAA-covered data in the cloud and consider yourself compliant with HIPAA regulations. You must ensure that your entire end-to-end computing infrastructure is compliant, and thus managing HIPAA data requires your institution's involvement and supervision.

One way to simplify the process of making a cloud-based computing infrastructure HIPAA compliant is to bring the cloud inside your institution's security boundary. This task can be accomplished in various ways by the cloud vendors. For example, Azure can create a special VPN that places a virtual secure partition of the Azure cloud directly into your network. Those cloud resources share your IP domain and can be accessed within your firewall. You will need your IT department to work with the cloud provider to set this up.

15.4 Secure Your VMs and Containers

You launch a VM or container on a cloud by using the methods described in chapter 4. What security threats do you need to be concerned about in this situation, and what should you do to overcome them? We have already talked about the need to protect the access key that you use to authenticate to your cloud provider when creating VM instances or containers. We are concerned here with what happens after that point. Figure 15.2 shows important activities performed when using VMs, and associated security risks.

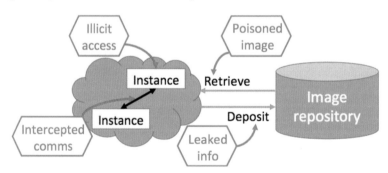

Figure 15.2: Hexagons showing four classes of risk associated with virtual machines, as images are retrieved from a repository; VM instances are created, execute, and communicate on a cloud; and modified images are added to a repository.

15.4.1 Poisoned VM or Container Image

Any time that you run a VM image or launch a container that you did not create yourself, the danger exists that the container has unwanted code that may, for example, make your private data accessible to others, participate in illicit activities such as denial of service attacks on other computers, or corrupt your computational results. Another concern is that a downloaded VM image may not be up to date with security patches and thus is vulnerable to attacks.

We overcome these concerns in much the same way as when installing software on a personal computer: we verify its source, ensure that it is up to date with any patches, and run it within a secure environment.

Verification of the source of a VM or container requires understanding the provenance of the image. In the case of VMs, each cloud vendor supplies a collection of trusted images that you can deploy; the cloud vendor also provides free malware tools that you may install once your image is running. In the case of containers, another solution is to provide a secure hash along with the image. This is a key that can be used to verify that the container image has not been tampered with. You can then use the hash key as part of the `docker pull` command. Mouat [206] provides an excellent overview of Docker security.

15.4.2 Illicit Access to Running VMs

Once created, a VM instance is little different from a computer running in your home, laboratory, or institutional data center. Thus, the risks that you face when running a VM on a cloud are essentially the same as those that you encounter when working on a physical computer. So too are the controls that you should deploy for protection. An important difference is that in the cloud case, you have more responsibility for implementing the necessary controls. Lest that last statement appear daunting, we point out that public cloud providers provide tools that make it easy to do the right thing. The following are important steps.

- *Limit who can access the instance.* You need to limit portal access, as anyone with account management access can change instance properties. If you wish to grant access only to the VM, then go to the VM, add the user with `sudo adduser` (assuming it is Linux), and add their public key to the newly created user's `.ssh` directory.

- *Ensure that the credentials that allow access to the instance are not compromised.* Thus, for example, create the instance with a key pair (see section 5.2.1 on page 75), and make sure that the private key is well protected.

- *Ensure that the software running on an instance is up to date with all security patches*, bearing in mind that a VM image downloaded from a repository may require updating before running, as noted above.

- If you run web applications such as Jupyter or a web service, make sure the network ports that they use are open and the software listening on those ports is not subject to known exploits. Run Jupyter with a key pair and password, as described in section 6.2. If you want more than one user to have access to Jupyter, it is better to run the JupyterHub multiuser system `jupyterhub.readthedocs.io`, rather than sharing one Jupyter instance and password, because Jupyter can open a shell on your account.

15.4.3 Intercepted Communications

The best way to secure the virtual services that you manage in the cloud is to remove them from the Internet by placing them in a **virtual private network** (VPN). A VPN is a layer on top of an existing network defined by point-to-point encrypted tunnels or a set of routes through a software defined network that carry encrypted packets. A VPN carries its own IP addresses and subnets that are not recognized as being part of the Internet.

You can set up a VPN in a number of ways; the choice depends on the network you need to create. For example, suppose you have a set of virtual machines in the cloud that you want to be on a private network that includes only those VMs and your laptop. Or, perhaps you have a private network in your lab or company that you want to extend to include your cloud resources. Or you may want some of your web services to be open to the public cloud, but you want those services also to be able to connect via a VPN to other servers not visible to the Internet. Each public cloud allows you to use their cloud portal to create a VPN that solves your specific problem. While the details of setting up a VPN are beyond the scope of this book, each cloud vendor provides extensive tutorials to guide you through the process. In the case of OpenStack the process is similar, depending on the specific OpenStack deployment.

15.4.4 Information Leakage via VM Image

We all commonly share images with our colleagues, either directly or via public repositories. As when pushing code to GitHub, you need to make sure that the images that you share do not contain credentials or other confidential information. It is even more important if you modify a public image and then push that image to an image repository for others to use. Bugiel et al. [82] tested 1,100 Amazon AMI EC2 images from Europe and the U.S. and found that about one-third contained an SSH back door: a public key that allows remote access to the instance. They were able to use this back door to extract AWS API keys, private keys, and credentials, as well as private data from many instances. Amazon warns users about this problem when Amazon discovers it, but you are well advised, whenever you clone an image from any repository, to look for any `authorize_keys` files in user home directories and delete them.

15.5   Secure Access to Cloud Software Services

The role-based access control systems described above define how you can delegate access and use of cloud-provided servers to your users, applications, and containers. However, this mechanism does not address the issue of how to control access to the services that you create for others and host in the cloud. These external users are not the people on your team that built the service and that have authorization to use your cloud account. These users are "customers" of the service, and you want to authorize them individually.

In developing such mechanisms, SSL and HTTPS are certainly important, as are passwords. You can create access control lists that can add some protection if you have a way to authenticate your users. Another solution to the authentication problem is to use a third-party authentication system. We have all seen online services that allow us to login using our Facebook or Google identity and password. The Azure app service provides a simple tool that you can use to enable Facebook, Google, or Microsoft as the authentication provider for your service.

The Globus tools can also help with authentication and authorization. We described in chapter 11 how Globus Auth can be used to authenticate with a variety of identity providers, including many university authentication systems, and how access tokens can be used for authorization. We also described how the Globus Auth SDK can be used to develop services and clients that apply these mechanisms. The Globus Genomics system described in chapter 14 illustrates the use of these mechanisms.

15.6   Summary

Cloud security is a major concern for many users, as it should be. It is also a complex topic that touches on so many cloud capabilities. In this chapter, we have presented only a light overview of the topic. If you are using the cloud, the first important issue is controlling who can access your data. Two types of access are involved here: people on your team to whom you have granted cloud accounts and external people with whom you want to share your data. Different mechanisms handle each case. IAM roles are good for the former case, and access keys and secure signatures are good for the latter. Globus Sharing is another powerful tool for sharing data securely with collaborators.

The second issue that we addressed is controlling access to your VMs and containers. Major issues to consider here are poisoned images, illicit access, intercepted communication, and information leakage. These concerns are handled with a variety of mechanisms. For poisoned images, the key is ascertaining the provenance of the image, deciding whether you trust it, and installing all provided malware protection. For illicit access, you must make sure that the software that you are running is free of exploits and that you manage access through SSH by adding the correct public keys. If you are concerned about intercepted communications, then you need to consider using a VPN. Information leakage is similar to the illicit access problem. A common problem is SSH back doors on VMs that can be easily used to compromise your VM. You need to check the `authorized_keys` file in the `.ssh` directory to make sure it is clean.

We also considered the problem of sharing secrets among containers. We described how Amazon's IAM role system, Azure's RBAC, and Docker's swarm secret sharing systems make such sharing possible. Moreover, we described how you can address security in the services that you create and expose to the world.

15.7 Resources

The *NIST Cloud Computing Security Reference Architecture* [97] provides an extensive review of cloud security issues and approaches. The Cloud Security Alliance `cloudsecurityalliance.org` produces and collects much relevant material on technologies and best practices. Chen et al. [93] and Hashizume et al. [151] provide useful reviews of issues. Amazon [254] and Azure [242] review best practices for VM security management on their systems. Huang et al. [157] survey academic research relating to IaaS security.

Illustrating the complexity of the security compliance environment, Amazon lists at `aws.amazon.com/compliance` more than 50 information sources and documents relating to certification and attestation programs in which they participate; laws, regulations, and privacy policies that apply in different contexts and countries; and security related frameworks. Few of these rules apply to most work on the cloud, but this breadth of material emphasizes the importance of getting expert guidance when working with sensitive data, whether in the cloud or elsewhere.

Chapter 16

History, Critiques, Futures

"I've looked at clouds from both sides now / From up and down and still somehow / It's cloud's illusions I recall / I really don't know clouds at all"

—Joni Mitchell

We have devoted 14 chapters to how you can use the cloud for scientific research. We now spend some time on context, covering, in turn, the historical context from which today's cloud emerged; contemporary critiques of cloud computing; and some important directions in which cloud technologies are developing. This material is brief, but we hope will stimulate thought and discussion.

16.1 Historical Perspectives

The idea of computing as a utility is far from new. Artificial intelligence pioneer Professor John McCarthy, speaking at MIT's centennial celebration in 1961, opined that: "Computing may someday be organized as a public utility just as the telephone system is a public utility." He went on to predict a future in which:

> "Each subscriber needs to pay only for the capacity he actually uses, but he has access to all programming languages characteristic of a very large system ... Certain subscribers might offer service to other subscribers ... The computer utility could become the basis of a new and important industry."

McCarthy's words were inspired by what he saw as the possibilities of time sharing, recently demonstrated in project Multics [100]. If many people could run on the same computer at the same time, then why not leverage economies of scale and use one computer to serve the needs of many people? This concept led to the mainframe, but it seems that McCarthy had something more ambitious in mind: perhaps a single computing utility to serve an entire nation? (At a similar talk at Stanford, McCarthy was apparently challenged by a physicist who observed that "this idea will never work: a simple back-of-the-envelope calculation shows that the amount of copper wire required to connect users to the computing utility would be impossible." This exchange provides a useful warning of the difficulties inherent in technological predictions, when new developments—in this case optical fiber—can upend fundamental assumptions. But it was also accurate: the large-scale realization of computing utilities was for a long time hindered by network limitations.)

These ideas continued to percolate in the imaginations of researchers. In 1966, Parkhill produced a prescient book-length analysis [217] of the challenges and opportunities of utility computing, and in 1969, when UCLA turned on the first node of the ARPANET, Internet pioneer Leonard Kleinrock claimed that "as [computer networks] grow up and become more sophisticated, we will probably see the spread of 'computer utilities' which, like present electric and telephone utilities, will service individual homes and offices across the country" [248].

The large-scale realization of computing utilities had to wait until networks were faster. In the early 1990s, various groups started to deploy then-new optical networking technologies for research purposes. In the US, **gigabit testbeds** linked a number of universities and research laboratories. Inspired by what might be possible now that computers were connected at speeds close to the memory bandwidth, researchers started to talk about **metacomputers** [237]—virtual computational systems created by linking components at different sites. Out of these discussions grew the idea of a computational grid, which "by analogy to the electric power grid provides access to power on demand, achieves economies of scale by aggregation of supply, and depends on large-scale federation of many suppliers and consumers for its effective operation" [126]. Software and protocols were developed for remote access to storage and computing, and many scientific communities leveraged these developments to federate computing facilities on local, national, and even global scales. For example, high energy physicists designing the Large Hadron Collider (LHC) realized that they needed to federate computing systems at hundreds of sites if they were to analyze the many petabytes of data to be produced by LHC experiments; in response, they developed the LHC Computing Grid (LCG) [175].

Grid computing enabled on-demand access to computing, storage, and other services, but its impact was primarily limited to science [127]. (One exception was within the enterprise, where "enterprise Grids" were widely deployed. These deployments are today often called "private clouds," with the principal difference being the use of virtualization to facilitate dynamic resource provisioning.) The emergence of cloud computing around 2006 is a fascinating story of marketing, business model, and technological innovation. A cynic could observe, with some degree of truth, that many articles from the 1990s and early 2000s on grid computing could be—and often were—republished by replacing every occurrence of "grid" with "cloud." But this is more a comment on the fashion- and hype-driven nature of technology journalism (and, we fear, much academic research in computer science) than on cloud itself. In practice, cloud is about the effective realization of the economies of scale to which early grid work aspired but did not achieve because of inadequate supply and demand. The success of cloud is due to profound transformations in these and other aspects of the computing ecosystem.

Cloud is driven, first and foremost, by a transformation in demand. It is no accident that the first successful infrastructure-as-a-service business emerged from an e-commerce provider. As Amazon CTO Werner Vogels tells the story, Amazon realized, after its first dramatic expansion, that it was building out literally hundreds of similar work-unit computing systems to support the different services that contributed to Amazon's online e-commerce platform. Each such system needed to be able to scale rapidly its capacity to queue requests, store data, and acquire computers for data processing. Refactoring across the different services produced services like Amazon's Simple Queue Service, Simple Storage Service, and Elastic Computing Cloud. Those services (and other similar services from other cloud providers, as described in previous chapters) have in turn been successful in the marketplace because many other e-commerce businesses need similar capabilities, whether to host simple e-commerce sites or to provide more sophisticated services such as video on demand.

Cloud is also enabled by a transformation in transmission. While the U.S. and Europe still lag behind broadband leaders such as South Korea and Japan, the number of households with megabits per second or faster connections is large and growing. One consequence is the widespread adoption of data-intensive services such as YouTube and Netflix. Another is that businesses feel increasingly able to outsource business processes such as email, customer relationship management, and accounting to software-as-a-service (SaaS) vendors.

Finally, cloud is enabled by a transformation in supply. Both IaaS vendors and companies offering consumer-facing services (e.g., search: Google, auctions: eBay, social networking: Facebook, Twitter) require enormous quantities of computing and storage. Leveraging advances in commodity computer technologies, these and other companies have learned how to meet those needs cost effectively within enormous data centers themselves [69] or, alternatively, have outsourced this aspect of their business to IaaS vendors. The commoditization of virtualization [67, 227] has facilitated this transformation, making it far easier than before to allocate computing resources on demand, with a precisely defined software stack installed.

In our opinion, it is the transformational changes in demand, transmission, and supply, and the resulting virtuous circle of increased use, better networks, and reduced costs, that account for the tremendous success of cloud technologies. It will be interesting to see where the next set of disruptive changes will occur, a topic that we consider in section 16.3.

16.2   Critiques

The reader will by now have realized that we are great fans of the power of the outsourcing and automation that cloud computing provides. We believe that by enabling users facing mundane or challenging computational tasks to focus on their problem, rather than the task of acquiring and operating computational infrastructure, cloud computing can frequently increase productivity and thus discovery and innovation.

Nevertheless it is also important to be aware of the various critiques that have been levied against cloud, some of which, in our opinion, speak to real or potential limitations and some to misunderstandings or differences of opinion. We review some of those critiques in the following. (As we have already discussed security concerns in chapter 15, we do not revisit them here.)

16.2.1   Cost

A common critique of cloud is that it is too expensive. We do not dismiss the importance of such concerns, particularly in academic settings where personnel and equipment spending may not be fungible. But without getting into the details of cost comparisons between on-premises and commercial cloud providers, we point out that when performing such comparisons, it is important to consider all costs, including personnel, space, and power. See, for example, Burt Holzman's 2016 analysis of in-house vs. public cloud computing costs for high energy physics [155]. He found that when power, cooling, and staff costs were included, on-premises computing in the Fermilab data center cost 0.9 cents per core hour under the assumption of 100% utilization, while off-premises computing on Amazon cost 1.4 cents per core hour. The observed computational speeds for their application were close to identical. Experience suggests that, depending on the specifics of your institutional computing environment and workload, cloud costs can be insignificant, greater than local costs, or less than local costs.

16.2.2 Lock In

Free software evangelist Richard Stallman [162] has argued that cloud computing is "simply a trap aimed at forcing more people to buy into locked, proprietary systems that [will] cost them more and more over time." He expands upon this point in an article in the Boston Review [238].

This is a common critique of cloud computing. At issue is the risk that arises when we become dependent for our computing on a third party provider. What if that provider goes out of business, discontinues services on which we depend, fails to meet desired quality of service commitments, or raises prices? What if they lose your data? These are real risks that any potential cloud user needs to evaluate, balancing them against the benefits that cloud brings. One partial hedge is to use only services for which equivalents exist from other providers, and to develop applications that use those services so that they can easily be retargeted. One way to do this is to build applications in containers, such as Docker, that allow them to run without modification on any commercial cloud. However, if the application in the container invokes a special platform service, such as a cloud-specific NoSQL service or stream broker, then changes are required. Good design and encapsulation of these dependencies in microservices can mitigate this problem. A more fundamental issue is the data stored in the cloud. Moving data can be difficult if they are large. The best solution may be to maintain an archive of the data elsewhere.

16.2.3 Education

We have heard people critique the use of cloud computing in education on the basis that students who rely on cloud services for storage and computing will not gain the hands-on knowledge that is gained from, for example, installing and operating Linux on a computer cluster. (We have both been asked variants of this rather disturbing question: "How will graduate students gain employment if they cannot perform systems administration tasks?")

It is easy to dismiss such concerns as Luddite misunderstandings of new technologies, but we feel that an important point is being made. One should rejoice in the capabilities that cloud computing provides, but we are the poorer if in seizing those benefits we lose understanding of the technologies that we are using. We should be educating students not simply how to use simple cloud services to perform simple tasks, but how cloud can be a platform for new approaches to science. We hope that this book can help in that process.

16.2.4 Black Box Algorithms

Another critique of cloud computing concerns the impact of handing off various aspects of your work to proprietary software developed and operated by third parties. If one cannot read the source code for a software component, obtain accurate documentation of the methods that it uses, or even test it comprehensively, then one has presumably lost the ability to determine the precise provenance of any results obtained with that software [205]. A related concern is that software on which one depends may be updated by a cloud provider without one's knowledge, in ways that turn out to affect your results.

These concerns appear to us to be quite real in the case of, for example, proprietary machine learning, data analytics, or computational modeling packages operated by cloud providers: in such cases, the result of a computation may indeed depend on decisions, changes, or errors made deep within complex software packages. We see fewer concerns in the case of systems software: while we may lack knowledge of how exactly a cloud provider implements a particular data management function, for example, the range of people using the software is larger and thus undetected errors are less likely.

These concerns are by no means new to cloud computing: they arise whenever results derive from software that cannot easily be studied or understood. (Microsoft Excel, for example, while simple to use, is a complex black box.) The high complexity and frequent updates associated with cloud software packages do arguably raise new challenges, but we suggest that simple approaches can be adopted. Use signals such as peer opinion and documentation to evaluate software quality. Test with problems for which you know the answers. Use cloud services for which source code is available—as it often is, as we have detailed in other chapters. In the case of machine learning methods, seek methods that yield models that are interpretable by human readers, so that the implications of a model can be understood and reviewed for hidden biases.

16.2.5 Hardware Limitations

A common critique of cloud computing, at least in the early days, was that cloud provided only limited hardware choices: it was ok if you wanted a vanilla x86 box, but not if you wanted something special. Today, the range of available hardware options is surely far greater than exists in any laboratory. Amazon, Azure and Google provide dozens of machine types, with varying quantities of CPU cores, memory, GPU capabilities, and other capabilities as we have described in section 7.2.2 on page 98.

16.3 Futures

The cloud that we have described in this book is the cloud of 2017. We believe that many of the technologies and the principles presented here will have a long life, but we also know that cloud technologies are evolving with great rapidity. (Amazon, Azure and Google all regularly announce dozens of new services and capabilities.) Thus we spend some time prognosticating about areas in which we believe cloud computing is likely to evolve in the next several years.

16.3.1 Cloud-native Applications

What does it mean to develop an application for the cloud? As we saw in chapter 4, it is straightforward to take many existing applications, package them so that they run in a virtual machine, and deploy that virtual machine onto a cloud compute service. But in so doing, all you have done is eliminate (or at least shift) hardware costs. You have not changed the essential nature of your applications in ways that take advantage of cloud features such as fault tolerant storage, elasticity, and powerful services such as those described in part III.

The term **cloud native** is used to describe applications that are written to take advantage of the powerful collections of services provided by cloud platforms. The **Cloud Native Computing Foundation** `www.cncf.io` writes that: "[c]loud native computing [deploys] applications as microservices, packaging each part into its own container, and dynamically orchestrating those containers to optimize resource utilization." They describe open source software packages available on Amazon, Azure, and Google, such as Kubernetes and Prometheus, that can be used to support such applications. We described microservice architecture in section 7.6 on page 110 and illustrated it with our simple scientific document analyzer. The cloud-native concept is more than just microservice implementations [120]. Cloud-native applications have a clear separation between persistent state, such as a database, and logic that runs in ephemeral virtual machines or containers, as shown in figure 16.1. The Globus service described in section 14.5 on page 307 has these characteristics.

The tools that you use to deploy such applications (Kubernetes, Mesos, etc.) also allow you to easily monitor and manage them. Such applications scale effortlessly and can be partitioned so that new versions of an application's microservices can be deployed and tested alongside the current "active" deployment. If the new versions work as planned, the old versions can be scaled back and no interruption in external service is seen.

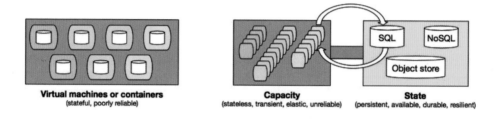

Virtual machines or containers
(stateful, poorly reliable)

Capacity
(stateless, transient, elastic, unreliable)

State
(persistent, available, durable, resilient)

Figure 16.1: On the left, the conventional deployment approach: each application is deployed in a virtual machine or container that contains all application state. On the right, the cloud-native approach: state is maintained in cloud data services and computation is performed by ephemeral service instances.

As we discussed in section 4.3 on page 67, serverless computing is about having the cloud manage collections of your functions to be executed on special conditions you define. This concept is closely related to cloud-native design. Unlike traditional scientific computations which run from start to finish, cloud-native apps run until you scale their implementation back to zero. Even in that quiescent state, they can be restarted simply by telling the deployment tool to increase from zero. One can set it so that an external event can trigger a serverless responder such as Amazon Lambda to invoke the deployment system to scale up the application.

So what does cloud-native have to do with the future of science? Consider the following scenario. Suppose you have a network of experimental instruments that produce data in large volumes and bursts, which you need to analyze as they arrive in real-time. This application can naturally be structured as a set of interactive microservices. One microservice receives and scans data. If something interesting is spotted, it invokes other microservices to perform additional processing, each of which may need to scale up to take on these tasks. These various components all send results to cataloging microservices that push data to a persistent data repository. A second category of events may be triggered by users making queries concerning the data that has been gathered. These other events can also cause other analysis tasks to be performed, or may just involve access to the data repository. The resulting cloud-native experiment management system may have dozens of individual microservice types, all interacting and scaling according to demand.

16.3.2 Architectural Evolution

Once upon a time, cloud data centers were built with racks of off-the-shelf servers from companies like Dell and HP. The relentless economics of competing in the cloud marketplace has completely changed the way data centers are designed. The first thing to go was off-the-shelf servers. Google was early in moving to cheap blade servers packed densely into racks. Amazon followed this practice and was soon building its own servers in collaboration with companies like Taiwan's Quanta. Traditional servers were just too expensive.

Big changes came around 2005 when those building massive data centers were forced to confront the fact that energy consumption was a major cost of doing business. Amazon, Google and Microsoft were experimenting with a variety of ideas to reduce the energy footprint of their data centers. This included tapping into renewable sources of energy such as geothermal, wind and wave action. Data center designs began to adopt supercomputer-style hot-cold aisle air conditioning. Microsoft was able to move to a system in which 2000 servers were packaged into a large shipping container that could be deployed outside.

The next phase of design evolution involved the server and not just its packaging and cooling. By 2010, many data cloud vendors were designing their own servers. In 2011, Facebook started the **Open Compute Project** [35] to create an open source design for the server itself. Facebook, Google, and Microsoft also began experimenting with ARM processors as a lower power alternative to the traditional Intel processor. As it became clear that different cloud workloads required different resources, the variety of server configurations began to explode.

The original data center designs used conventional commercial networking gear at the top of each rack and between racks. As these centers grew, their networking needs became more demanding. Institutions demanded ways to extend their private network directly into the cloud through scalable, virtual private networks. By 2012, the Azure network was all based on software defined networks [228]; the same is true for Amazon and Google.

The most recent architectural changes in the cloud are being driven by the performance requirement of search, analytics, and machine learning. In 2010. Microsoft research began a study of how to optimize the Bing search algorithms. This work evolved into a major redesign of server architecture around **Field Programmable Gate Arrays** (FPGAs) that have been added to the Azure servers [85]. The FPGAs are situated between the network switches and the servers so that this programmable logic lies in a plane allowing FPGA-to-FPGA direct communication. This architecture, called Catapult, allows applications needing special acceleration to group together a set of FPGAs and servers into a special purpose mesh. This configuration is used for applications like high speed encryption and accelerating deep learning [216]. Microsoft is not the only cloud that is deploying custom hardware. Google recently announced the Tensor Processing Unit [164], which is designed to be a better accelerator for TensorFlow than GPUs.

These examples of cloud data center evolution illustrate that the designs are moving rapidly toward a possible convergence with supercomputer technology. While the cloud will always have a different use model than the largest supercomputers, we expect the value of the cloud for science to only increase.

16.3.3 Edge Computing

Cloud computing has become synonymous with massive, hyper-connected data centers, within which storage and computation are allocated fluidly in response to user demand. This highly centralized architecture has been central to cloud computing's success, permitting both economies of scale in terms of operations costs and innovative applications that depend on the aggregation and analysis of large quantities of data. And as cloud provider services continue to increase in sophistication, and as businesses, homes, and people become increasingly well connected, it can easily seem that there is no limit to the applications that can be moved from personal computers to the cloud. Perhaps, we may think, all computing will soon occur elsewhere.

Yet at the same time as cloud data centers become more powerful and people become more connected to those data centers, other important trends are pushing towards decentralization. Increasingly powerful sensors generate vast quantities of data that often cannot be cost effectively transferred to cloud data centers but must be processed locally. Increasing demands for computer-in-the-loop control make latency increasingly critical. Consider, for example, an automated observation system that is to detect migrating birds and then zoom in to obtain high-resolution images that can be used to identify individual animals. It is likely not practical to stream real-time video from thousands of cameras to the cloud, process the data, and return results in time to zoom the cameras. But an inexpensive local processing unit, perhaps running algorithms configured based on large-scale offline machine learning, can easily perform such tasks.

For applications such as these, computing needs to occur "at the edge" of the network: hence **edge computing** [232]; the term "fog computing," another nebulous neologism, is sometimes also used [75]. Of course, that is where computing has always been performed, at least since the PC era. But a new question being considered is how the edge and the cloud may be connected. Will we see cloud providers start to engineer cloud services that extend out to the edge? What will this mean for what we choose to outsource to the cloud? It will be fascinating to see how these questions are answered over the next decade and beyond.

We can already see early examples of cloud providers extending the reach of their services beyond their primary data centers. **Content distribution networks** (e.g., Akamai, Amazon CloudFront, Azure CDN) run edge servers distributed worldwide (68 such servers for Amazon CloudFront, as of 2017) to cache content (e.g., web pages) that is to be made available rapidly to clients. More intriguing are developments in serverless computing. As we saw in section 4.3 on page 67, services such as Amazon Lambda, Azure Functions, and Google Cloud Functions allow users to define functions to be performed when certain events occur. While these services make it possible to implement powerful reactive applications, their responsiveness will be limited if every event notification and subsequent response have to travel from the origin site to a cloud data center. Thus, Amazon provides **Lambda"@Edge**, which allows functions to run on Amazon CloudFront content delivery network nodes. Intriguingly, they have also announced plans to allow Lambda functions to "execute on hardware that isn't a part of Amazon's cloud or doesn't have a consistent connection to the internet" [132]: perhaps, for example, on computers associated with experimental apparatus in a scientific laboratory or on Internet of Things components such as the Array of Things nodes described in section 9.1.2 on page 163.

16.4   Resources

The History of the Grid [126] reviews many developments relevant to utility, grid, and cloud computing.

Chapter 17

Jupyter Notebooks

"He listens well who takes notes"
—Dante, *The Divine Comedy, Canto XV*

We provide accompanying Jupyter notebooks to illustrate the use of various technologies described in this book. These notebooks explain selected techniques and approaches and provide thorough implementation details so that you can quickly start using the technologies covered within. They combine explanation, basic exercises, and substantial additional Python code to provide a conceptual understanding of each technology, give insight into how key parts of the process are implemented through exercises, and then lay out an end-to-end pattern for implementing each in your own work. The notebooks are interactive documents that mix formatted text and Python code samples that can be edited and run in real-time in a Jupyter notebook server, allowing you to run and explore the code for each technology as you read about it.

The notebooks and related files are accessible online at `Cloud4SciEng.org`. The **Notebooks** menu tab contains the list of notebooks provided below, with links to both HTML renderings and the source `.ipynb` files. The **Extras** tab contains links to other non-notebook source code referred to in the book. They are freely available to be downloaded by anyone at any time, and run on any appropriately configured computer. In most cases, additional packages need to be added; each notebook includes instructions for adding any such required packages.

17.1 Environment

You need a version of Python and Jupyter installed on your computer to run the notebooks, including a local Jupyter server that you will use to run the notebooks. You also need to install additional Python packages needed by the notebooks, and a few additional programs. The easiest way to get these components installed and working is to install the free Anaconda Python distribution provided by Continuum Analytics `continuum.io/downloads`. Anaconda includes a Jupyter server and precompiled versions of many of the packages used in the notebooks. Alternatively, you can also create your Python environment manually, installing Python, package managers, and Python packages separately. Packages like NumPy and Pandas can be difficult to get working, however, particularly on Windows. Anaconda simplifies this setup considerably, regardless of your operating system. We discuss elsewhere in the book how to configure virtual machines and containers to run Python and Jupyter in the cloud.

17.2 The Notebooks

We provide brief descriptions for each of the 23 notebooks. As the reader is surely well aware, software rusts. Versions of APIs and SDKs are replaced by newer versions. We will try our best to keep these notebooks rust-free.

Notebook 1 provides a first look at Jupyter. It illustrates Jupyter features that we use extensively, including the mixing of text and LaTeX math with Python code, and the use of inline graphics.

Notebook 2 provides the first of four implementations of a scenario described in section 3.1 on page 37. You have a CSV file describing some experimental data and a collection of that data. The task is to create a table in the cloud, upload the data for each experiment to blob storage, and then add to the table a row containing the metadata for that experiment and a URL for the associated data. This notebook uses Amazon DynamoDB for the table and the S3 storage service for the blobs.

Notebook 3 implements the same scenario as notebook 2, but using the Azure table and blob services.

Notebook 4 provides a partial implementation of the scenario, using Google Bigtable to create a table.

Notebook 5 completes the scenario, using the Google Datastore and blob storage.

Notebook 6 illustrates the use of the CloudBridge Python package to manage basic storage operations on the OpenStack layer of the Jetstream cloud.

Notebook 7 shows how to create and manage virtual machines on Amazon using the Boto3 Python library.

Notebook 8 illustrates how Python can be used to manage file transfers and share data with Globus.

Notebook 9 describes how to use the Amazon EC2 Container Service, as described in section 7.6.4 on page 114. It shows how to interact with the container service and launch new containers. You will find additional data files as well as the Docker files needed to build the containers at the **EXTRAS** tab of the book website. Notebook 10 also is a part of this project.

Notebook 10 implements a client program that feeds data into a queue, to be consumed by microservices from notebook 9.

Notebook 11 uses Spark to perform a trivial MapReduce computation.

Notebook 12 provides a second demonstration of Spark, in this case for a k-means clustering algorithm.

Notebook 13 illustrates the use of a special set of commands that allow us to embed SQL in a Jupyter notebook directly.

Notebook 14 shows how to deploy Jupyter in a Spark cluster on an Amazon Elastic Map Reduce cluster, using the exploration of Wikipedia data as an illustrative example.

Notebook 15 uses Google's Datalab to explore contagious disease records from the U.S. Centers for Disease Control, specifically looking at Rubella cases over a period of time.

Notebook 16 further applies Google's Datalab, using it to examine weather station data and to identify an anomaly in one station's reporting.

Notebook 17 uses Amazon Kinesis together with Spark to detect anomalies in data from Array of Things instrument streams. The data needed for this project can be found at the **EXTRAS** tab of the book website.

Notebook 18 uses Azure's HDInsight plus Spark to look at food inspection records.

Notebook 19 implements a client that can be used to push data to a web service described in section 10.2 on page 197: a simple document classifier built with the Azure ML tool.

Notebook 20 shows how to reconstitute a recurrent neural network created by training CNTK with text from business news items, and then load and run the resulting model.

Notebook 21 is based on the MXNet example of section 10.5 on page 212, in which MXNet was used to train the Resnet-152 image recognition model. It loads and runs the trained network to identify images from the web.

Notebook 22 is a description of how to deploy CNTK on your local machine.

Notebook 23 uses TensorFlow to build a simple logistic regression analyzer that can be used to make simple predictions of graduate school admissions.

17.3 Resources

There are many ways to get Jupyter running. The easiest is to install Anaconda and then run the command `Jupyter notebook`. This works on your PC, Mac or on a virtual machine in the cloud. In the case of a VM in the cloud you can use a container or directly download Anaconda. You will need to create a `.jupyter` directory and follow the instruction in section 10.5 on page 212 where it describes the installation for MXNET on AWS.

In various parts of this book we have provided instructions on other ways to run Jupyter. For example, to run it in a container, see section 6.2 on page 87; with Google's Kubernetes, see section 7.6.5 on page 120; and with Spark in a container, see section 8.2.3 on page 141. We noted in section 1.5 on page 12 the value of Python and Git/GitHub for scientists and engineers who intend to work with cloud technologies.

Chapter 18

Afterword: A Discovery Cloud

"It would appear that we have reached the limits of what it is possible to achieve with computer technology, although one should be careful with such statements, as they tend to sound pretty silly in five years."

—John von Neumann

We hope that the preceding pages have given you some concrete ideas about how you can use the cloud in your research. Perhaps the cloud, for you, will simply be a place to store your research data securely and cheaply, or to perform computations that you could not easily run before. Or perhaps you are inspired to embrace the power of the cloud to transform how you run your laboratory, conduct your research, and interact with your community. No matter how you approach the use of these technologies, we are confident that you will find the experience both rewarding and fun.

We cannot resist this last opportunity to prognosticate. The pioneering cybernetician and organizational theorist Stafford Beer wrote in 1972 [70]:

> The question which asks how to use the computer in the enterprise [is] the wrong question. A better formulation is to ask how the enterprise should be run given that computers exist. The best version of all is the question asking what, given computers, the enterprise now is.

We, in turn, are fascinated by the following variant of Beer's best question:

> Given cloud, and all that its scalable and cost-effective automation and outsourcing entail, what the scientific enterprise now is.

We propose the following likely outcomes. **Industrialization of data production** via large-scale automated experiments and observations, already occurring in astronomy [241], functional genomics, and materials science [159], will expand to many more domains. The resulting data glut will in turn drive **industrialization of data analysis**, by which we mean large-scale computational platforms that automate quality control, analysis, inference, and other steps. These developments will greatly reduce the costs of hypothesis generation and testing. They will also improve reproducibility because experimental configurations and data processing steps will be captured precisely.

Meanwhile, the digital encoding of large quantities of scientific knowledge from such experiments and other sources (e.g., the scientific literature) will enable the creation of a **universal knowledge base** supporting both rapid access and automated inference. It will become routine to ask questions via a scientific search engine, to be notified of potential inconsistencies across existing knowledge, and to vote on the next set of experiments to be performed by industrial-scale facilities. Other experiments will be performed by quasi-independent **robot scientists** [171, 262] that apply inference and experiment design methods to guide their choice of the next experiment.

These steps towards economies of scale may sound dehumanizing, but experience suggests that, if implemented in the right way, they can unleash a flood of creativity. If the universal knowledge base is treated as a globally accessible public good, then the scientific playing field becomes more level. A high school student in Angola, India, or New Zealand will be able to search for new drugs for rare diseases or new materials sourced from local materials. They will use powerful tools accessible from this **discovery cloud** [124] to collect new data, analyze extant and new data, test hypotheses, and contribute to knowledge.

This discovery cloud will also empower the bench scientist. Amazon's Echo and Alexa services can keep our calendar, order a pizza, and summon a car service, and Alexa can invoke an Amazon Lambda function to start an experimental analysis in the cloud. All of these actions can be driven by voice commands. As machine learning continues to progress, future scientists will benefit from a **cloud-based research assistant** that not only monitors experiments but also performs background research, such as scanning the literature for related work and checking our mathematical derivations. Such a system will respond to vocal instructions while also reading (and writing) our computational notebooks.

Some of these developments may be some way out, but the technology is evolving fast. As Roy Amara observed, "[w]e tend to overestimate the effect of a technology in the short run and underestimate the effect in the long run."

Bibliography

[1] Access control at the project level. https://cloud.google.com/storage/docs/access-control/iam.

[2] Apache Flink dataflow programming model. https://ci.apache.org/projects/flink/flink-docs-release-1.2/concepts/programming-model.html.

[3] Assignments for Udacity deep learning class with TensorFlow. https://github.com/tensorflow/tensorflow/tree/master/tensorflow/examples/udacity.

[4] AWS Case Study: Animoto. https://aws.amazon.com/solutions/case-studies/animoto/.

[5] AWS Identity and Access Management best practices. http://docs.aws.amazon.com/IAM/latest/UserGuide/best-practices.html.

[6] Azure Batch Shipyard recipes. https://github.com/Azure/batch-shipyard/tree/master/recipes.

[7] Azure Data Lake Store Python SDK. https://github.com/Azure/azure-data-lake-store-python.

[8] Azure: Deploy a slurm cluster. https://github.com/Azure/azure-quickstart-templates/tree/master/slurm/README.md.

[9] Bare metal on OpenStack: Ironic. https://wiki.openstack.org/wiki/Ironic.

[10] CentOS 7 / RHEL 7 – Open ports. http://www.linuxbrigade.com/centos-7-rhel-7-open-ports/.

[11] Cloudbridge documentation. http://cloudbridge.readthedocs.io/en/latest/.

[12] Containers on OpenStack: Magnum. https://wiki.openstack.org/wiki/Magnum.

[13] Deep learning AMI Amazon Linux version. https://aws.amazon.com/marketplace/pp/B01M0AXXQB.

[14] Euca2ools overview. `https://docs.hpcloud.com/eucalyptus/4.3.0/euca2ools-guide/index.html`.

[15] Eucalyptus EDGE network configuration. `https://docs.eucalyptus.com/eucalyptus/4.3/install-guide/nw_edge.html`.

[16] Eucalyptus installation guide. `https://docs.eucalyptus.com/eucalyptus/latest/shared/install_section.html`.

[17] Eucalyptus network configuration requirements. `https://docs.hpcloud.com/eucalyptus/4.3.0/install-guide/preparing_firewalls.html`.

[18] Eucalyptus: Plan services placement. `https://docs.eucalyptus.com/eucalyptus/latest/install-guide/services_understanding.html`.

[19] Eucalyptus: Planning networking modes. `https://docs.eucalyptus.com/eucalyptus/latest/install-guide/planning_networking_modes.html`.

[20] Galaxy on Jetstream. `https://wiki.galaxyproject.org/Cloud/Jetstream`.

[21] Get started: Create Apache Spark cluster on HDInsight Linux and run interactive queries using Spark SQL. `https://azure.microsoft.com/en-us/documentation/articles/hdinsight-apache-spark-jupyter-spark-sql/`.

[22] Globus endpoint activation. `https://docs.globus.org/api/transfer/endpoint_activation/`.

[23] Google Cloud Dataflow: Complete Examples. `http:https://cloud.google.com/dataflow/examples/all-examples`.

[24] Google Cloud Datalab Quickstart. `https://cloud.google.com/datalab/docs/quickstarts/quickstart-local`.

[25] IBM Analytics Stream Computing. `http://www.ibm.com/analytics/us/en/technology/stream-computing/`.

[26] The Kubernetes project. `http://kubernetes.io`.

[27] Layers library reference. `https://www.cntk.ai/pythondocs/layerref.html`.

[28] Linux RAID. `https://raid.wiki.kernel.org/index.php/Linux_Raid`.

[29] Machine Learning Library (MLlib) guide. `https://spark.apache.org/docs/latest/ml-guide.html`.

[30] Making secure requests to Amazon Web Services. `https://aws.amazon.com/articles/1928`.

[31] Microsoft Azure Event Hubs. `https://azure.microsoft.com/en-us/services/event-hubs/`.

[32] Microsoft Azure Stack. `https://azure.microsoft.com/en-us/overview/azure-stack/`.

[33] NCBI BLAST on Windows Azure. `https://www.microsoft.com/en-us/download/details.aspx?id=52513`.

[34] Ocean Observatories Initiative. `http://oceanobservatories.org`.

[35] The Open Compute Project. `http://opencompute.org`.

[36] OpenStack documentation: CPU topologies. `https://docs.openstack.org/admin-guide/compute-cpu-topologies.html`.

[37] OpenStack in production: Hints and tips from the CERN OpenStack cloud team. `http://openstack-in-production.blogspot.co.uk/`.

[38] OpenStack Newton release notes. `https://www.openstack.org/software/newton/`.

[39] OpenStack: Operators mailing list. `http://lists.openstack.org/pipermail/openstack-operators/`.

[40] OpenStack: Scientific working group. `https://wiki.openstack.org/wiki/Scientific_working_group`.

[41] Predict with pre-trained models. `http://mxnet.io/tutorials/python/predict_imagenet.html`.

[42] Rados object storage utility. `http://docs.ceph.com/docs/giant/man/8/rados/`.

[43] Riak cloud storage. `http://docs.basho.com/riak/cs/2.1.1/`.

[44] Sample applications built using Amazon Machine Learning. `https://github.com/awslabs/machine-learning-samples`.

[45] Spark SQL, DataFrames and Datasets guide. `http://spark.apache.org/docs/latest/sql-programming-guide.html`.

[46] The Red Hat Package Manager. `https://en.wikipedia.org/wiki/RPM_Package_Manager`.

[47] Theano deep learning library. `http://www.deeplearning.net/software/theano/`.

[48] Transferring RDA data with Globus. `http://ncarrda.blogspot.com/2015/06/transferring-rda-data-with-globus.html`.

[49] TripleO online documentation. `http://tripleo.org`.

[50] VMware Cloud Foundation. `https://www.vmware.com/products/cloud-foundation.html`.

[51] Welcome to Bridges. `https://www.psc.edu/index.php/resources/computing/bridges`.

[52] What is IAM? `https://docs.aws.amazon.com/IAM/latest/UserGuide/Introduction.html`.

[53] Setup Linux Network Bridges on CentOS for Nova Networking, Nov 2015. `https://platform9.com/support/setup-network-bridges-on-centos-nova-networking/`.

[54] OpenStack user survey, Oct 2016. `https://www.openstack.org/assets/survey/October2016SurveyReport.pdf`.

[55] Using AWS in the context of New Zealand privacy considerations. Technical report, Oct. 2016. `https://d0.awsstatic.com/whitepapers/compliance/Using_AWS_in_the_context_of_New_Zealand_Privacy_Considerations.pdf`.

[56] G. Agha. An overview of actor languages. *SIGPLAN Notices*, 21(10):58–67, June 1986.

[57] T. Akidau. The world beyond batch: Streaming 102, Jan 2016. `https://www.oreilly.com/ideas/the-world-beyond-batch-streaming-102`.

[58] T. Akidau, R. Bradshaw, C. Chambers, S. Chernyak, R. J. Fernández-Moctezuma, R. Lax, S. McVeety, D. Mills, F. Perry, E. Schmidt, and S. Whittle. The dataflow model: A practical approach to balancing correctness, latency, and cost in massive-scale, unbounded, out-of-order data processing. *Proceedings of the VLDB Endowment*, 8(12):1792–1803, Aug. 2015.

[59] T. Akidau and F. Perry. Dataflow/Beam and Spark: A Programming Model Comparison, Feb 2016. `https://cloud.google.com/dataflow/blog/dataflow-beam-and-spark-comparison`.

[60] A. Aliper, S. Plis, A. Artemov, A. Ulloa, P. Mamoshina, and A. Zhavoronkov. Deep learning applications for predicting pharmacological properties of drugs and drug repurposing using transcriptomic data. *Molecular Pharmaceutics*, 2016.

[61] W. Allcock, J. Bresnahan, R. Kettimuthu, M. Link, C. Dumitrescu, I. Raicu, and I. Foster. The Globus striped GridFTP framework and server. In *ACM/IEEE Conference on Supercomputing*, page 54, 2005.

[62] B. Allen, J. Bresnahan, L. Childers, I. Foster, G. Kandaswamy, R. Kettimuthu, J. Kordas, M. Link, S. Martin, K. Pickett, and S. Tuecke. Software as a service for data scientists. *Communications of the ACM*, 55(2):81–88, Feb. 2012.

[63] S. Anthony. How big is the Cloud? *ExtremeTech*, May 2012. `http://www.extremetech.com/computing/129183-how-big-is-the-cloud`.

[64] M. Armbrust, R. S. Xin, C. Lian, Y. Huai, D. Liu, J. K. Bradley, X. Meng, T. Kaftan, M. J. Franklin, A. Ghodsi, et al. Spark SQL: Relational data processing in Spark. In *ACM SIGMOD International Conference on Management of Data*, pages 1383–1394, 2015.

[65] P. Bailis and K. Kingsbury. The network is reliable. *Queue*, 12(7):20, 2014.

[66] R. Barga, J. Goldstein, M. Ali, and M. Hong. Consistent streaming through time: A vision for event stream processing. In *Conference on Innovative Data Systems Research*, pages 363–374, 2007.

[67] P. Barham, B. Dragovic, K. Fraser, S. Hand, T. Harris, A. Ho, R. Neugebauer, I. Pratt, and A. Warfield. Xen and the art of virtualization. *ACM SIGOPS Operating Systems Review*, 37(5):164–177, 2003.

[68] W. Barnett, V. Welch, A. Walsh, and C. A. Stewart. A roadmap for using NSF cyberinfrastructure with InCommon, 2011. `https://www.incommon.org/federation/cyberroadmap.html`.

[69] L. A. Barroso, J. Clidaras, and U. Hölzle. The datacenter as a computer: An introduction to the design of warehouse-scale machines. *Synthesis Lectures on Computer Architecture*, 8(3):1–154, 2013.

[70] S. Beer. *Brain of the Firm*. Penguin Press, 1972.

[71] D. Bernstein. Containers and cloud: From LXC to Docker to Kubernetes. *IEEE Cloud Computing*, 1(3):81–84, 2014.

[72] P. Bernstein, S. Berkov, J. Thelin, and S. Burkhardt. Orleans - Virtual Actors. `http://research.microsoft.com/en-us/projects/orleans/`.

[73] K. Bhuvaneshwar, D. Sulakhe, R. Gauba, A. Rodriguez, R. Madduri, U. Dave, L. Lacinski, I. Foster, Y. Gusev, and S. Madhavan. A case study for cloud based high throughput analysis of NGS data using the Globus Genomics system. *Computational and Structural Biotechnology Journal*, 13:64–74, 2015.

[74] M. Bojarski, D. Del Testa, D. Dworakowski, B. Firner, B. Flepp, P. Goyal, L. D. Jackel, M. Monfort, U. Muller, J. Zhang, X. Zhang, and J. Zhao. End to end learning for self-driving cars. *arXiv preprint arXiv:1604.07316*, 2016.

[75] F. Bonomi, R. Milito, J. Zhu, and S. Addepalli. Fog computing and its role in the internet of things. In *MCC Workshop on Mobile Cloud Computing*, pages 13–16. ACM, 2012.

[76] D. E. Boyle, D. C. Yates, and E. M. Yeatman. Urban sensor data streams: London 2013. *IEEE Internet Computing*, 17(6):12–20, 2013.

[77] T. Bray. One Amazon year, December 2015. `https://www.tbray.org/ongoing/When/201x/2015/12/01/One-Amazon-Year`.

[78] E. Brewer. CAP twelve years later: How the "rules" have changed. *Computer*, 45(2):23–29, 2012.

[79] E. Brewer. Kubernetes and the path to cloud native. In *6th ACM Symposium on Cloud Computing*, pages 167–167. ACM, 2015.

[80] J. Bryce. Embracing datacenter diversity. In *OpenStack Austin*. 2016. `https://www.openstack.org/videos/video/embracing-datacenter-diversity`.

[81] Y. Bu, B. Howe, M. Balazinska, and M. D. Ernst. HaLoop: Efficient iterative data processing on large clusters. *Proceedings of the VLDB Endowment*, 3(1-2):285–296, 2010.

[82] S. Bugiel, S. Nürnberger, T. Pöppelmann, A.-R. Sadeghi, and T. Schneider. AmazonIA: When elasticity snaps back. In *18th ACM conference on Computer and Communications Security*, pages 389–400. ACM, 2011.

[83] J. Cantarella, C. Shonkwiler, and E. Uehara. A fast direct sampling algorithm for equilateral closed polygons, Jan 2017. `https://arxiv.org/abs/1510.02466v2`.

[84] C. Catlett, T. Malik, B. Goldstein, J. Giuffrida, Y. Shao, A. Panella, D. Eder, E. v. Zanten, R. Mitchum, S. Thaler, and I. Foster. Plenario: An open data discovery and exploration platform for urban science. *Bulletin of the IEEE Computer Society Technical Committee on Data Engineering*, pages 27–42, 2014.

[85] A. Caulfield, E. Chung, A. Putnam, H. Angepat, J. Fowers, M. Haselman, S. Heil, M. Humphrey, P. Kaur, J.-Y. Kim, D. Lo, T. Massengill, K. Ovtcharov, M. Papamichael, L. Woods, S. Lanka, D. Chiou, and D. Burger. A cloud-scale acceleration architecture. In *49th Annual IEEE/ACM International Symposium on Microarchitecture*, October 2016.

[86] M. Cezar. Setting up NTP (Network Time Protocol) Server in RHEL/CentOS 7. *Tecmint*, Mar 2015. `http://www.tecmint.com/install-ntp-server-in-centos/`.

[87] K. M. Chandy, O. Etzion, and R. von Ammon. Event processing. Dagstuhl Seminar Proceedings 10201, Schloss Dagstuhl - Leibniz-Zentrum fuer Informatik, Germany, 2011.

[88] K. Chard, S. Caton, O. F. Rana, and K. Bubendorfer. Social cloud: Cloud computing in social networks. *IEEE CLOUD*, 10:99–106, 2010.

[89] K. Chard, J. Pruyne, B. Blaiszik, R. Ananthakrishnan, S. Tuecke, and I. Foster. Globus data publication as a service: Lowering barriers to reproducible science. In *11th IEEE International Conference on eScience*, pages 401–410, 2015.

[90] R. Chard, K. Chard, K. Bubendorfer, L. Lacinski, R. Madduri, and I. Foster. Cost-aware cloud provisioning. In *IEEE 11th International Conference on e-Science*, pages 136–144, 2015.

[91] R. Chard, R. Madduri, N. Karonis, K. Chard, K. Duffin, C. Ordonez, T. Uram, J. Fleischauer, I. Foster, M. Papka, and J. Winans. Scalable pCT image reconstruction delivered as a cloud service. *IEEE Transactions on Cloud Computing*, 2015. http://ieeexplore.ieee.org/document/7160740/.

[92] T. Chen, M. Li, Y. Li, M. Lin, N. Wang, M. Wang, T. Xiao, B. Xu, C. Zhang, and Z. Zhang. Mxnet: A flexible and efficient machine learning library for heterogeneous distributed systems. *CoRR*, abs/1512.01274, 2015.

[93] Y. Chen, V. Paxson, and R. H. Katz. What's new about cloud computing security. *University of California, Berkeley Report No. UCB/EECS-2010-5 January*, 2010.

[94] T. Cheng and J. Wang. *Application of a Dynamic Recurrent Neural Network in Spatio-Temporal Forecasting*, pages 173–186. Springer Berlin Heidelberg, Berlin, Heidelberg, 2007.

[95] K. Cho, B. Van Merriënboer, D. Bahdanau, and Y. Bengio. On the properties of neural machine translation: Encoder-decoder approaches. *arXiv preprint arXiv:1409.1259*, 2014.

[96] J. Clark. 5 numbers that illustrate the mind bending size of Amazon's cloud. *Bloomberg Global Tech*, Nov 2014. http://www.bloomberg.com/news/2014-11-14/5-numbers-that-illustrate-the-mind-bending-size-of-amazon-s-cloud.html.

[97] Cloud Computing Security Working Group. NIST Cloud Computing Security Reference Architecture. Special Publication 500-299, National Institute of Standards and Technology, 2013. http://collaborate.nist.gov/twiki-cloud-computing/pub/CloudComputing/CloudSecurity/NIST_Security_Reference_Architecture_2013.05.15_v1.0.pdf.

[98] D. T. Cohen, G. W. Hatchard, and S. G. Wilson. Population trends in incorporated places: 2000 to 2013. Technical Report P25-1142, US Census, Mar 2015.

[99] A. Conesa, P. Madrigal, S. Tarazona, D. Gomez-Cabrero, A. Cervera, A. McPherson, M. W. Szcześniak, D. J. Gaffney, L. L. Elo, X. Zhang, et al. A survey of best practices for RNA-seq data analysis. *Genome biology*, 17(1):13, 2016.

[100] F. J. Corbató and V. Vyssotsky. Introduction and overview of the Multics system. *IEEE Annals of the History of Computing*, 14(2):12–13, 1992.

[101] J. C. Corbett, J. Dean, M. Epstein, A. Fikes, C. Frost, J. J. Furman, S. Ghemawat, A. Gubarev, C. Heiser, P. Hochschild, W. Hsieh, S. Kanthak, E. Kogan, H. Li, A. Lloyd, S. Melnik, D. Mwaura, D. Nagle, S. Quinlan, R. Rao, L. Rolig, Y. Saito, M. Szymaniak, C. Taylor, R. Wang, , and D. Woodford. Spanner: Google's globally distributed database. *ACM Transactions on Computer Systems*, 31(3):8, 2013.

[102] T. Cowles, J. Delaney, J. Orcutt, and R. Weller. The Ocean Observatories Initiative: Sustained ocean observing across a range of spatial scales. *Marine Technology Society Journal*, 44(6):54–64, 2010.

[103] D. R. Cox. The regression analysis of binary sequences. *Journal of the Royal Statistical Society. Series B (Methodological)*, pages 215–242, 1958.

[104] R. J. Creasy. The origin of the VM/370 time-sharing system. *IBM Journal of Research and Development*, 25(5):483–490, 1981.

[105] J. Czyzyk, M. P. Mesnier, and J. J. Moré. The NEOS server. *IEEE Computational Science and Engineering*, 5(3):68–75, 1998.

[106] E. Dart, L. Rotman, B. Tierney, M. Hester, and J. Zurawski. The Science DMZ: A network design pattern for data-intensive science. *Scientific Programming*, 22(2):173–185, 2014.

[107] F. De Carlo. DMagic data management system. `http://dmagic.readthedocs.io`.

[108] J. Dean and S. Ghemawat. MapReduce: Simplified data processing on large clusters. *Communications of the ACM*, 51(1):107–113, 2008.

[109] E. Deelman, K. Vahi, M. Rynge, G. Juve, R. Mayani, and R. F. da Silva. Pegasus in the cloud: Science automation through workflow technologies. *IEEE Internet Computing*, 20(1):70–76, 2016.

[110] P. Dhingra, K. Tolle, and D. Gannon. Using cloud-based analytics to save lives. *Cloud Computing in Ocean and Atmospheric Sciences*, page 221, 2016.

[111] S. Dieleman. My solution for the Galaxy Zoo challenge, Apr 2014. `http://benanne.github.io/2014/04/05/galaxy-zoo.html`.

[112] C. Docan, M. Parashar, and S. Klasky. DataSpaces: An interaction and coordination framework for coupled simulation workflows. *Cluster Computing*, 15(2):163–181, 2012.

[113] A. Dubey and D. Wagle. Delivering software as a service. *The McKinsey Quarterly*, 6(2007):2007, 2007.

[114] D. Eadline. *Hadoop 2 Quick-Start Guide: Learn the Essentials of Big Data Computing in the Apache Hadoop 2 Ecosystem*. Addison-Wesley, 2016.

[115] G. Eisenhauer, M. Wolf, H. Abbasi, and K. Schwan. Event-based systems: Opportunities and challenges at exascale. In *3rd ACM International Conference on Distributed Event-Based Systems*, 2009.

[116] S. Ekanayake, S. Kamburugamuve, and G. Fox. SPIDAL: high performance data analytics with Java and MPI on large multicore HPC clusters. In *Spring Simulation Multi-Conference*, pages 3–6, 2016.

[117] J. Elliott, D. Kelly, J. Chryssanthacopoulos, M. Glotter, K. Jhunjhnuwala, N. Best, M. Wilde, and I. Foster. The parallel system for integrating impact models and sectors (pSIMS). *Environmental Modelling & Software*, 62:509–516, 2014.

[118] O. Etzioni. Deep learning isn't a dangerous magic genie. It's just math. *Wired*, June 2016. `https://www.wired.com/2016/06/deep-learning-isnt-dangerous-magic-genie-just-math/`.

[119] B. Familiar. *Microservices, IoT and Azure: Leveraging DevOps and Microservice Architecture to deliver SaaS Solutions*. APress, 2015.

[120] M. R. Ferré. Cloud native applications (for dummies), 2014. `http://www.it20.info/2014/12/cloud-native-applications-for-dummies/`.

[121] R. T. Fielding. *Architectural styles and the design of network-based software architectures*. PhD thesis, University of California, Irvine, 2000.

[122] J. Fischer, S. Tuecke, I. Foster, and C. A. Stewart. Jetstream: A distributed cloud infrastructure for underresourced higher education communities. In *1st Workshop on The Science of Cyberinfrastructure: Research, Experience, Applications and Models*, pages 53–61. ACM, 2015.

[123] I. Foster. Globus Online: Accelerating and democratizing science through cloud-based services. *IEEE Internet Computing*, 15(3):70–73, May 2011.

[124] I. Foster, K. Chard, and S. Tuecke. The discovery cloud: Accelerating and democratizing research on a global scale. In *IEEE International Conference on Cloud Engineering*, pages 68–77. IEEE, 2016.

[125] I. Foster, R. Ghani, R. S. Jarmin, F. Kreuter, and J. I. Lane, editors. *Big Data and Social Science: A Practical Guide to Methods and Tools*. Taylor & Francis Group, 2016. See also `http://www.bigdatasocialscience.com`.

[126] I. Foster and C. Kesselman. The history of the grid. In *High Performance Computing: From Grids and Clouds to Exascale*, pages 3–30. IOS Press, 2011.

[127] I. Foster, Y. Zhao, I. Raicu, and S. Lu. Cloud computing and grid computing 360-degree compared. In *Grid Computing Environments Workshop*, pages 1–10. IEEE, 2008.

[128] A. Fox, D. A. Patterson, and S. Joseph. *Engineering Software as a Service: An Agile Approach using Cloud Computing*. Strawberry Canyon LLC, 2013.

[129] G. Fox and D. Gannon. Using clouds for technical computing, 2013. `http://www.academia.edu/14845479/Using_Clouds_for_Technical_Computing`.

[130] G. Fox, S. Jha, and L. Ramakrishnan. Streaming and Steering Applications: Requirements and Infrastructure. `http://streamingsystems.org`.

[131] G. C. Fox, R. D. Williams, and G. C. Messina. *Parallel Computing Works!* Morgan Kaufmann, 2014.

[132] B. H. Frank. AWS wants to dominate beyond the public cloud with Lambda updates. *PC World*, Dec. 2016. http://www.pcworld.com/article/3147389/.

[133] D. Gannon. Performance Analysis of a Cloud Microservice-based ML Classifier, Oct 2015. https://esciencegroup.com/2015/10/08/performance-analysis-of-a-cloud-microservice-based-ml-classifier/.

[134] D. Gannon. CNTK revisited. A new deep learning toolkit release from Microsoft, Nov 2016. https://esciencegroup.com/2016/11/10/cntk-revisited-a-new-deep-learning-toolkit-release-from-microsoft/.

[135] D. Gannon, D. Fay, D. Green, K. Takeda, and W. Yi. Science in the cloud: Lessons from three years of research projects on Microsoft Azure. In *5th International Workshop on Scientific Cloud Computing*, pages 1–8. ACM, 2014.

[136] Gartner Research. Software as a Service (SaaS). http://www.gartner.com/it-glossary/software-as-a-service-saas.

[137] K. Gee and W. Hunt. Enhancing stormwater management benefits of rainwater harvesting via innovative technologies. *Journal of Environmental Engineering*, 142(8):04016039, 2016.

[138] L. George. *HBase: The Definitive Guide: Random Access to Your Planet-Size Data.* O'Reilly Media, Inc., 2011.

[139] A. V. Gerbessiotis and L. G. Valiant. Direct bulk-synchronous parallel algorithms. *Journal of Parallel and Distributed Computing*, 22(2):251–267, 1994.

[140] S. Goasguen. Enjoy Kubernetes with Python. https://www.linux.com/learn/kubernetes/enjoy-kubernetes-python.

[141] J. Goecks, A. Nekrutenko, J. Taylor, and T. G. Team. Galaxy: A comprehensive approach for supporting accessible, reproducible, and transparent computational research in the life sciences. *Genome Biol*, 11(8):R86, 2010.

[142] J. Gong, P. Yue, and H. Zhou. Geoprocessing in the Microsoft cloud computing platform–Azure. In *Joint Symposium of ISPRS Technical Commission IV & AutoCarto*, page 6, 2010.

[143] I. Goodfellow, Y. Bengio, and A. Courville. *Deep Learning*. MIT Press, 2016. http://www.deeplearningbook.org.

[144] A. Graves, A. Mohamed, and G. Hinton. Speech recognition with deep recurrent neural networks. In *IEEE International Conference on Acoustics, Speech, and Signal Processing*, pages 6645–6649. IEEE, 2013.

[145] A. Greenberg, J. Hamilton, D. A. Maltz, and P. Patel. The cost of a cloud: Research problems in data center networks. *ACM SIGCOMM Computer Communication Review*, 39(1):68–73, 2008.

[146] K. Gremban. Get started with access management in the Azure portal. `https://docs.microsoft.com/en-us/azure/active-directory/role-based-access-control-what-is`.

[147] W. Gropp, E. Lusk, and R. Thakur. *Using MPI-2: Advanced Features of the Message Passing Interface*. MIT Press, 1999.

[148] J. Han, M. Kamber, and J. Pei. *Data Mining: Concepts and Techniques*. Morgan-Kaufmann, 2011.

[149] D. Hardt. OAuth 2.0 authorization framework specification, 2012. `http://tools.ietf.org/html/rfc6749`.

[150] J. A. Hartigan and M. A. Wong. Algorithm AS 136: A k-means clustering algorithm. *Journal of the Royal Statistical Society. Series C (Applied Statistics)*, 28(1):100–108, 1979.

[151] K. Hashizume, D. G. Rosado, E. Fernández-Medina, and E. B. Fernandez. An analysis of security issues for cloud computing. *Journal of Internet Services and Applications*, 4(1):1, 2013.

[152] K. He, X. Zhang, S. Ren, and J. Sun. Deep residual learning for image recognition. *CoRR*, abs/1512.03385, 2015.

[153] T. Hey, S. Tansley, and K. Tolle. *The Fourth Paradigm: Data-Intensive Scientific Discovery*. Kindle, 2009.

[154] B. Hindman, A. Konwinski, M. Zaharia, A. Ghodsi, A. D. Joseph, R. H. Katz, S. Shenker, and I. Stoica. Mesos: A platform for fine-grained resource sharing in the data center. In *USENIX Symposium on Networked Systems Design and Implementation*, pages 22–22, 2011.

[155] B. Holzman. Fermilab HEPCloud: An elastic computing facility for High Energy Physics. In *International Conference on Computing in High Energy Physics*. 2016. `https://indico.cern.ch/event/432527/contributions/1072465/`.

[156] A. Howard. Running MPI applications in Amazon EC2, May 2015. `https://cyclecomputing.com/running-mpi-applications-in-amazon-ec2/`.

[157] W. Huang, A. Ganjali, B. H. Kim, S. Oh, and D. Lie. The state of public infrastructure-as-a-service cloud security. *ACM Computing Surveys*, 47(4):68, 2015.

[158] T. Hunt. Introducing "Have I been pwned?" – aggregating accounts across website breaches, Dec 2013. `https://www.troyhunt.com/introducing-have-i-been-pwned/`.

[159] A. Jain, S. P. Ong, G. Hautier, W. Chen, W. D. Richards, S. Dacek, S. Cholia, D. Gunter, D. Skinner, G. Ceder, et al. The materials project: A materials genome approach to accelerating materials innovation. *APL Materials*, 1(1):011002, 2013.

[160] S. Jain, A. Kumar, S. Mandal, J. Ong, L. Poutievski, A. Singh, S. Venkata, J. Wanderer, J. Zhou, M. Zhu, J. Zolla, U. Hölzle, S. Stuart, and A. Vahdat. B4: Experience with a globally-deployed software defined WAN. *ACM SIGCOMM Computer Communication Review*, 43(4):3–14, 2013.

[161] Y. Jia and E. Shelhamer. Caffe. http://caffe.berkeleyvision.org/.

[162] B. Johnson. Cloud computing is a trap, warns GNU founder Richard Stallman. *Guardian Newspaper*, Sep 2008. https://www.theguardian.com/technology/2008/sep/29/cloud.computing.richard.stallman.

[163] B. Jones. Towards the European open science cloud, 2015. http://doi.org/10.5281/zenodo.16001.

[164] N. Jouppi. Google supercharges machine learning tasks with TPU custom chip, 2016. https://cloudplatform.googleblog.com/2016/05/Google-supercharges-machine-learning-tasks-with-custom-chip.html.

[165] S. Kamburugamuve and G. Fox. Survey of distributed stream processing, Feb 2016. https://www.researchgate.net/publication/299411481.

[166] S. Kamburugamuve, P. Wickramasinghe, S. Ekanayake, and G. Fox. Anatomy of machine learning algorithm implementations in MPI, Spark, and Flink, Jan 2017. https://www.researchgate.net/publication/312426658.

[167] N. T. Karonis, K. L. Duffin, C. E. Ordoñez, B. Erdelyi, T. D. Uram, E. C. Olson, G. Coutrakon, and M. E. Papka. Distributed and hardware accelerated computing for clinical medical imaging using proton computed tomography (pCT). *Journal of Parallel and Distributed Computing*, 73(12):1605–1612, 2013.

[168] A. Karpathy. The unreasonable effectiveness of recurrent neural networks, Feb 2015. http://karpathy.github.io/2015/05/21/rnn-effectiveness.

[169] M. Kassner. A look at Amazon's world class data center ecosystem, Dec 2014. http://www.techrepublic.com/article/a-look-at-amazons-world-class-data-center-ecosystem.

[170] S. Kemp. Password-less logins with OpenSSH, 2005. https://debian-administration.org/article/152/.

[171] R. D. King, J. Rowland, S. G. Oliver, M. Young, W. Aubrey, E. Byrne, M. Liakata, M. Markham, P. Pir, L. N. Soldatova, A. Sparkes, K. E. Whelan, and A. Clare. The automation of science. *Science*, 324(5923):85–89, 2009.

[172] G. Klimeck, M. McLennan, S. P. Brophy, G. B. Adams III, and M. S. Lundstrom. nanohub.org: Advancing education and research in nanotechnology. *Computing in Science & Engineering*, 10(5):17–23, 2008.

[173] S. Kulkarni, N. Bhagat, M. Fu, V. Kedigehalli, C. Kellogg, S. Mittal, J. M. Patel, K. Ramasamy, and S. Taneja. Twitter Heron: Stream processing at scale. In *ACM SIGMOD International Conference on Management of Data*, pages 239–250, 2015.

[174] H. S. Kuyuk, R. M. Allen, H. Brown, M. Hellweg, I. Henson, and D. Neuhauser. Designing a network-based earthquake early warning algorithm for California: ElarmS-2. *Bulletin of the Seismological Society of America*, 2013.

[175] M. Lamanna. The LHC computing grid project at CERN. *Nuclear Instruments and Methods in Physics Research Section A: Accelerators, Spectrometers, Detectors and Associated Equipment*, 534(1):1–6, 2004.

[176] K. A. Lawrence, M. Zentner, N. Wilkins-Diehr, J. A. Wernert, M. Pierce, S. Marru, and S. Michael. Science gateways today and tomorrow: Positive perspectives of nearly 5000 members of the research community. *Concurrency and Computation: Practice and Experience*, 27(16):4252–4268, 2015.

[177] J. Layton. A container for HPC. `https://www.admin-magazine.com/HPC/Articles/Singularity-A-Container-for-HPC`.

[178] J. A. Le, H. El-Askary, M. Allali, and D. Struppa. Application of recurrent neural networks for drought projections in California. *Atmospheric Research*, 187, 2017.

[179] H. Lee. Simple Azure. `https://readthedocs.org/projects/simple-azure/`.

[180] P. D. Lena, K. Nagata, and P. F. Baldi. Deep spatio-temporal architectures and learning for protein structure prediction. In *Advances in Neural Information Processing Systems 25*, pages 512–520. Curran Associates, Inc., 2012.

[181] Y. Li. Introduction to Docker secrets management. `https://blog.docker.com/2017/02/docker-secrets-management`.

[182] D. Lifka, I. Foster, S. Mehringer, M. Parashar, P. Redfern, C. Stewart, and S. Tuecke. XSEDE cloud survey report, 2013. `http://hdl.handle.net/2142/45766`.

[183] I. Liu and B. Ramakrishnan. Bach in 2014: Music composition with recurrent neural network. *CoRR*, abs/1412.3191, 2014.

[184] Q. Liu, J. Logan, Y. Tian, H. Abbasi, N. Podhorszki, J. Y. Choi, S. Klasky, R. Tchoua, J. Lofstead, R. Oldfield, M. Parashar, N. Samatova, K. Schwan, A. Shoshani, M. Wolf, K. Wu, and W. Yu. Hello ADIOS: The challenges and lessons of developing leadership class I/O frameworks. *Concurrency and Computation: Practice and Experience*, 26(7):1453–1473, 2014.

[185] Y. Liu, A. Padmanabhan, and S. Wang. CyberGIS Gateway for enabling data-rich geospatial research and education. *Concurrency and Computation: Practice and Experience*, 27(2):395–407, 2015.

[186] R. Madduri, K. Chard, R. Chard, L. Lacinski, A. Rodriguez, D. Sulakhe, D. Kelly, U. Dave, and I. Foster. The Globus Galaxies platform: Delivering science gateways as a service. *Concurrency and Computation: Practice and Experience*, 27(16):4344–4360, 2015.

[187] R. Madduri, D. Sulakhe, L. Lacinski, B. Liu, A. Rodriguez, K. Chard, U. J. Dave, and I. T. Foster. Experiences building Globus Genomics: A next-generation sequencing analysis service using Galaxy, Globus, and Amazon Web Services. *Concurrency and Computation: Practice and Experience*, 26(13):2266–2279, 2014.

[188] P. K. Mantha, A. Luckow, and S. Jha. Pilot-MapReduce: An extensible and flexible MapReduce implementation for distributed data. In *3rd International Workshop on MapReduce and Its Applications*, pages 17–24. ACM, 2012.

[189] J. Margolis. Amazon Echo's role in deep space exploration. *Financial Times*, Jan 2017. https://www.ft.com/content/24529e30-d0e3-11e6-b06b-680c49b4b4c0.

[190] N. Marz and J. Warren. *Big Data – Principles and Best Practices of Scalable Realtime Data Systems*. Manning, 2015.

[191] A. Matsunaga, J. Fortes, K. Keahey, and M. Tsugawa. Sky computing. *IEEE Internet Computing*, 13:43–51, 2009.

[192] K. Matthias and S. P. Kane. *Docker: Up and Running*. O'Reilly, 2016.

[193] W. McKinney. *Python for Data Analysis: Data Wrangling with Pandas, NumPy, and IPython*. O'Reilly Media, 2015.

[194] N. Mehrotra, L. Franks, P. McKay, R. McAllister, and J. Gao. Get started: Create Apache Spark cluster in Azure HDInsight and run interactive queries using Spark SQL. https://docs.microsoft.com/en-us/azure/hdinsight/hdinsight-apache-spark-jupyter-spark-sql/.

[195] N. Mehrotra, R. McMurray, L. Franks, and J. Gao. Machine learning: Predictive analysis on food inspection data using MLlib with Apache Spark cluster on HDInsight Linux. https://docs.microsoft.com/en-us/azure/hdinsight/hdinsight-apache-spark-machine-learning-mllib-ipython.

[196] P. Mehrotra, J. Djomehri, S. Heistand, R. Hood, H. Jin, A. Lazanoff, S. Saini, and R. Biswas. Performance evaluation of Amazon EC2 for NASA HPC applications. In *3rd Workshop on Scientific Cloud Computing*, pages 41–50. ACM, 2012.

[197] P. Mell and T. Grance. The NIST definition of cloud computing. Special Publication 800-145, National Institute of Standards and Technology, 2011. http://nvlpubs.nist.gov/nistpubs/Legacy/SP/nistspecialpublication800-145.pdf.

[198] X. Meng, J. Bradley, B. Yavuz, E. Sparks, S. Venkataraman, D. Liu, J. Freeman, D. Tsai, M. Amde, S. Owen, D. Xin, R. Xin, M. J. Franklin, R. Zadeh, M. Zaharia, and A. Talwalkary. MLlib: Machine learning in Apache Spark. *Journal of Machine Learning Research*, 17(34):1–7, 2016.

[199] F. Meyer, D. Paarmann, M. D'Souza, R. Olson, E. M. Glass, M. Kubal, T. Paczian, A. Rodriguez, R. Stevens, A. Wilke, et al. The metagenomics RAST server–A public resource for the automatic phylogenetic and functional analysis of metagenomes. *BMC bioinformatics*, 9(1):386, 2008.

[200] Microsoft Research Connections. MSR Courseware. `https://github.com/MSRConnections/Azure-training-course`.

[201] M. A. Miller, W. Pfeiffer, and T. Schwartz. Creating the CIPRES science gateway for inference of large phylogenetic trees. In *Gateway Computing Environments Workshop*, pages 1–8, 2010.

[202] D. Milojičić, I. M. Llorente, and R. S. Montero. OpenNebula: A cloud management tool. *IEEE Internet Computing*, 15(2):11–14, 2011.

[203] N. M. Mohamed, H. Lin, and W.-C. Feng. Accelerating data-intensive genome analysis in the cloud. In *5th International Conference on Bioinformatics and Computational Biology*. 2013.

[204] T. P. Morgan. A rare peek at the massive scale of AWS. *EnterpriseTech*, Nov 2014. `http://www.enterprisetech.com/2014/11/14/rare-peek-massive-scale-aws`.

[205] A. Morin, J. Urban, P. D. Adams, I. Foster, A. Sali, D. Baker, and P. Sliz. Shining light into black boxes. *Science*, 336(6078):159–160, 2012.

[206] A. Mouat. Docker security: Using containers safely in production. `https://gallery.mailchimp.com/979c70339150d05eec1531104/files/Docker_Security_Red_Hat.pdf`.

[207] A. C. Muller and S. Guido. *Introduction to Machine Learning with Python: A Guide for Data Scientists*. O'Reilly Publishing, 2017.

[208] N. Nakata, J. P. Chang, J. F. Lawrence, and P. Boué. Body wave extraction and tomography at long beach, california, with ambient-noise interferometry. *Journal of Geophysical Research: Solid Earth*, 120(2):1159–1173, 2015.

[209] F. Nelli. *Python Data Analytics: Data Analysis and Science using Pandas, Matplotlib and the Python Programming Language*. Apress, 2015.

[210] M. A. Nielsen. *Neural Networks and Deep Learning*. Determination Press, 2015. `http://neuralnetworksanddeeplearning.com`.

[211] B. Nikolic. Data processing for the Square Kilometre Array telescope. `http://www.mrao.cam.ac.uk/~bn204/publications/2015/SKA-SDP-Streaming.pdf`.

[212] D. Nurmi, R. Wolski, C. Grzegorczyk, G. Obertelli, S. Soman, L. Youseff, and D. Zagorodnov. The Eucalyptus open-source cloud-computing system. In *9th IEEE/ACM International Symposium on Cluster Computing and the Grid*, pages 124–131, 2009.

[213] C. Olah. Understanding LSTM networks, Aug 2015. `http://colah.github.io/posts/2015-08-Understanding-LSTMs/`.

[214] C. Olston, B. Reed, U. Srivastava, R. Kumar, and A. Tomkins. Pig latin: A not-so-foreign language for data processing. In *ACM SIGMOD International Conference on Management of Data*, pages 1099–1110, 2008.

[215] R. Orihuela and D. Bass. Help wanted: Black belts in data, Jun 2015. `http://www.bloomberg.com/news/articles/2015-06-04/help-wanted-black-belts-in-data`.

[216] K. Ovtcharov, O. Ruwase, J.-Y. Kim, J. Fowers, K. Strauss, and E. Chung. Toward accelerating deep learning at scale using specialized hardware in the datacenter. In *27th HotChips Symposium on High-Performance Chips*. IEEE, August 2015.

[217] D. F. Parkhill. *The Challenge of the Computer Utility*. Addison-Wesley Educational Publishers, 1966.

[218] N. Paskin. Digital object identifier (DOI) system. *Encyclopedia of Library and Information Sciences*, 3:1586–1592, 2010.

[219] F. Pérez and B. Granger. The state of Jupyter, Jan 2017. `https://www.oreilly.com/ideas/the-state-of-jupyter`.

[220] D. A. Phillips, C. Puskas, Santillan, L. M., Wang, R. W. King, W. M. Szeliga, T. Melbourne, M. Murray, M. Floyd, and T. A. Herring. Plate Boundary Observatory and related networks: GPS data analysis methods and geodetic products. *Reviews of Geophysics*, 54:759–f808, 2016.

[221] I. Raicu, I. Foster, and Y. Zhao. Many-task computing for grids and supercomputers. In *IEEE Workshop on Many-Task Computing on Grids and Supercomputers*, 2008.

[222] K. Ram. Git can facilitate greater reproducibility and increased transparency in science. *Source Code for Biology and Medicine*, 8(1):7, 2013.

[223] L. Ramakrishnan, P. T. Zbiegel, S. Campbell, R. Bradshaw, R. S. Canon, S. Coghlan, I. Sakrejda, N. Desai, T. Declerck, and A. Liu. Magellan: Experiences from a science cloud. In *2nd International Workshop on Scientific Cloud Computing*, pages 49–58. ACM, 2011.

[224] S. Rashka. *Python Machine Learning*. Packt Publishing, 2016.

[225] K. Reitz. Requests: HTTP for humans. `http://docs.python-requests.org`.

[226] J. Richer. OAuth 2.0 token introspection. RFC 7662, IETF, 2015.

[227] M. Rosenblum and T. Garfinkel. Virtual machine monitors: Current technology and future trends. *Computer*, 38(5):39–47, 2005.

[228] M. Russinovich. Report from Open Networking Summit: Achieving hyper-scale with software defined networking. `http://bit.ly/21aCxLT`.

[229] S. Ryza, U. Laserson, S. Owen, and J. Wills. *Advanced Analytics with Spark: Patterns for Learning from Data at Scale*. O'Reilly Media, 2015.

[230] N. Sakimura, J. Bradley, M. Jones, B. d. Medeiros, and C. Mortimore. OpenID Connect Core 1.0 incorporating errata set 1, 2014. `http://openid.net/specs/openid-connect-core-1_0.html`.

[231] D. Sanderson. *Programming Google App Engine with Python: Build and Run Scalable Python Apps on Google's Infrastructure*. O'Reilly Press, 2015.

[232] M. Satyanarayanan. The emergence of edge computing. *Computer*, 50(1):30–39, 2017.

[233] C. Severance. Python for informatics: Exploring information, 2013. `http://www.pythonlearn.com/book.php`.

[234] D. Silver, A. Huang, C. J. Maddison, A. Guez, L. Sifre, G. Van Den Driessche, J. Schrittwieser, I. Antonoglou, V. Panneershelvam, M. Lanctot, S. Dieleman, D. Grewe, J. Nham, N. Kalchbrenner, I. Sutskever, T. Lillicrap, M. Leach, K. Kavukcuoglu, T. Graepel, and D. Hassabis. Mastering the game of Go with deep neural networks and tree search. *Nature*, 529(7587):484–489, 2016.

[235] F. Simorjay. Shared responsibilities for cloud computing. Technical report, Microsoft, Mar 2016. `https://gallery.technet.microsoft.com/Shared-Responsibilities-81d0ff91`.

[236] A. Singh, J. Ong, A. Agarwal, G. Anderson, A. Armistead, R. Bannon, S. Boving, G. Desai, B. Felderman, P. Germano, et al. Jupiter rising: A decade of Clos topologies and centralized control in Google's datacenter network. *ACM SIGCOMM Computer Communication Review*, 45(4):183–197, 2015.

[237] L. Smarr and C. E. Catlett. Metacomputing. *Communications of the ACM*, 35(6):44–53, 1992.

[238] R. M. Stallman. Who does that server really serve? *Boston Review*, 35(2), 2010.

[239] R. Stevens, P. Woodward, T. DeFanti, and C. Catlett. From the I-WAY to the National Technology Grid. *Communications of the ACM*, 40(11):50–60, 1997.

[240] C. Strasser. Git/GitHub: A primer for researchers, 2014. `http://datapub.cdlib.org/2014/05/05/github-a-primer-for-researchers/`.

[241] A. Szalay and J. Gray. The world-wide telescope. *Science*, 293(5537):2037–2040, 2001.

[242] T. Tetrick. Best practices for securing access to your Azure virtual machines, 2014. `https://blogs.technet.microsoft.com/uspartner_ts2team/2014/06/04/best-practices-for-securing-access-to-your-azure-virtual-machines/`.

[243] D. Thain, T. Tannenbaum, and M. Livny. Distributed computing in practice: The Condor experience. *Concurrency and Computation: Practice and Experience*, 17(2-4):323–356, 2005.

[244] B. Tierney, J. Metzger, J. Boote, E. Boyd, A. Brown, R. Carlson, M. Zekauskas, J. Zurawski, M. Swany, and M. Grigoriev. perfsonar: Instantiating a global network measurement framework. In *SOSP Workshop on Real Overlays and Distributed Systems*, 2009.

[245] J. Towns, T. Cockerill, M. Dahan, I. Foster, K. Gaither, A. Grimshaw, V. Hazlewood, S. Lathrop, D. Lifka, G. D. Peterson, R. Roskies, J. R. Scott, and N. Wilkins-Diehr. XSEDE: Accelerating scientific discovery. *Computing in Science & Engineering*, 16(5):62–74, 2014.

[246] R. Tudoran, A. Costan, G. Antoniu, and H. Soncu. TomusBlobs: Towards communication-efficient storage for MapReduce applications in Azure. In *12th IEEE/ACM International Symposium on Cluster, Cloud and Grid Computing*, pages 427–434, 2012.

[247] S. Tuecke, R. Ananthakrishnan, K. Chard, M. Lidman, B. McCollam, and I. Foster. Globus Auth: A research identity and access management platform. In *12th IEEE International Conference on e-Science*, 2016.

[248] T. Tugend. UCLA to be first station in nationwide computer network, July 1969. `http://www.lk.cs.ucla.edu/LK/Bib/REPORT/press.html`.

[249] J. Turnbull. *The Docker Book: Containerization is the New Virtualization*. Kindle, 2014.

[250] A. Vahdat. A look inside Google's data center networks, 2015. `https://cloudplatform.googleblog.com/2015/06/A-Look-Inside-Googles-Data-Center-Networks.html`.

[251] J. van Vliet and F. Paganelli. *Programming AWS EC2*. O'Reilly Press, 2011.

[252] T. C. Vance, N. Merati, C. Yang, and M. Yuan. *Cloud Computing in Ocean and Atmospheric Sciences.* Elsevier, 2016.

[253] J. VanderPlas. *Python Data Science Handbook: Essential Tools for Working with Data.* O'Reilly Media, 2017.

[254] J. Varia. Tips for securing your EC2 instance. `https://aws.amazon.com/articles/1233`.

[255] N. Vijayakumar and B. Plale. Performance evaluation of rate-based join window sizing for asynchronous data streams. In *13th IEEE International Symposium on High Performance Distributed Computing*, pages 260–261, 2004.

[256] W. Vogels. MXNet – Deep learning framework of choice at AWS, Nov 2016. `http://www.allthingsdistributed.com/2016/11/mxnet-default-framework-deep-learning-aws.html`.

[257] M. M. Waldrop. *The Dream Machine: JCR Licklider and the Revolution that Made Computing Personal.* Viking Penguin, 2001.

[258] T. White. *Hadoop: The Definitive Guide.* O'Reilly Media, Inc., 2012.

[259] M. Wilde, M. Hategan, J. M. Wozniak, B. Clifford, D. S. Katz, and I. Foster. Swift: A language for distributed parallel scripting. *Parallel Computing*, 37(9):633–652, 2011.

[260] J. Wilkening, A. Wilke, N. Desai, and F. Meyer. Using clouds for metagenomics: A case study. In *IEEE International Conference on Cluster Computing*, pages 1–6, 2009. `http://www.mcs.anl.gov/papers/P1665A.pdf`.

[261] N. Wilkins-Diehr, D. Gannon, G. Klimeck, S. Oster, and S. Pamidighantam. Tera-Grid science gateways and their impact on science. *Computer*, 41(11), 2008.

[262] K. Williams, E. Bilsland, A. Sparkes, W. Aubrey, M. Young, L. N. Soldatova, K. De Grave, J. Ramon, M. de Clare, W. Sirawaraporn, S. G. Oliver, and R. D. King. Cheaper faster drug development validated by the repositioning of drugs against neglected tropical diseases. *Journal of the Royal Society Interface*, 12(104):20141289, 2015.

[263] A. Wittig and M. Wittig. *Amazon Web Services in Action.* Manning Press, 2015.

[264] D. Xue, P. V. Balachandran, J. Hogden, J. Theiler, D. Xue, and T. Lookman. Accelerated search for materials with targeted properties by adaptive design. *Nature Communications*, 7, 2016.

[265] M. Zaharia, M. Chowdhury, M. J. Franklin, S. Shenker, and I. Stoica. Spark: Cluster computing with working sets. In *2nd USENIX Workshop on Hot Topics in Cloud Computing*, 2010. `https://www.usenix.org/legacy/event/hotcloud10/tech/full_papers/Zaharia.pdf`.

[266] Y. Zheng, X. Chen, Q. Jin, Y. Chen, X. Qu, X. Liu, E. Chang, W.-Y. Ma, Y. Rui, and W. Sun. A cloud-based knowledge discovery system for monitoring fine-grained air quality. Technical Report MSR-TR-2014–40, Microsoft Research, 2014.

Index

Scientific and Engineering Computation

William Gropp and Ewing Lusk, editors; Janusz Kowalik, founding editor

Data-Parallel Programming on MIMD Computers, Philip J. Hatcher and Michael J. Quinn, 1991

Enterprise Integration Modeling: Proceedings of the First International Conference, edited by Charles J. Petrie, Jr., 1992

The High Performance Fortran Handbook, Charles H. Koelbel, David B. Loveman, Robert S. Schreiber, Guy L. Steele Jr. and Mary E. Zosel, 1994

PVM: A User's Guide and Tutorial for Network Parallel Computing, Al Geist, Adam Beguelin, Jack Dongarra, Weicheng Jiang, Robert Manchek, and Vaidyalingham S. Sunderam, 1994

Practical Parallel Programming, Gregory V. Wilson, 1995

Enabling Technologies for Petaflops Computing, Thomas Sterling, Paul Messina, and Paul H. Smith, 1995

An Introduction to High-Performance Scientific Computing, Lloyd D. Fosdick, Elizabeth R. Jessup, Carolyn J. C. Schauble, and Gitta Domik, 1995

Parallel Programming Using C++, edited by Gregory V. Wilson and Paul Lu, 1996

Using PLAPACK: Parallel Linear Algebra Package, Robert A. van de Geijn, 1997

Fortran 95 Handbook, Jeanne C. Adams, Walter S. Brainerd, Jeanne T. Martin, Brian T. Smith, and Jerrold L. Wagener, 1997

MPI—The Complete Reference: Volume 1, The MPI Core, Marc Snir, Steve Otto, Steven Huss-Lederman, David Walker, and Jack Dongarra, 1998

MPI—The Complete Reference: Volume 2, The MPI-2 Extensions, William Gropp, Steven Huss-Lederman, Andrew Lumsdaine, Ewing Lusk, Bill Nitzberg, William Saphir, and Marc Snir, 1998

A Programmer's Guide to ZPL, Lawrence Snyder, 1999

How to Build a Beowulf, Thomas L. Sterling, John Salmon, Donald J. Becker, and Daniel F. Savarese, 1999

Using MPI-2: Advanced Features of the Message-Passing Interface, William Gropp, Ewing Lusk, and Rajeev Thakur, 1999

Beowulf Cluster Computing with Windows, edited by Thomas Sterling, William Gropp, and Ewing Lusk, 2001

Beowulf Cluster Computing with Linux, second edition, edited by Thomas Sterling, William Gropp, and Ewing Lusk, 2003

Scalable Input/Output: Achieving System Balance, edited by Daniel A. Reed, 2003

Using OpenMP: Portable Shared Memory Parallel Programming, Barbara Chapman, Gabriele Jost, and Ruud van der Pas, 2008

Quantum Computing without Magic: Devices, Zdzislaw Meglicki, 2008

Quantum Computing: A Gentle Introduction, Eleanor G. Rieffel and Wolfgang H. Polak, 2011

Using MPI: Portable Parallel Programming with the Message-Passing Interface, third edition, William Gropp, Ewing Lusk, and Anthony Skjellum, 2015

Using Advanced MPI: Beyond the Basics, Pavan Balaji, William Gropp, Torsten Hoefler, Rajeev Thakur, and Ewing Lusk, 2015

Scientific Programming and Computer Architecture, Divakar Viswanath, 2017

Cloud Computing for Science and Engineering, Ian Foster and Dennis B. Gannon, 2017